Principles and Techniques of Electron Microscopy

Principles and Techniques of Electron Microscopy

Biological Applications

Volume 4

Edited by

M. A. HAYAT

Professor of Biology
Kean College of New Jersey
Union, New Jersey

VNR **VAN NOSTRAND REINHOLD COMPANY**
New York/Cincinnati/Toronto/London/Melbourne

It is a pleasure to dedicate this volume to
Professor R. W. G. Wyckoff

Van Nostrand Reinhold Company Regional Offices:
New York   Cincinnati   Chicago   Millbrae   Dallas

Van Nostrand Reinhold Company International Offices:
London   Toronto   Melbourne

Library of Congress Catalog Card Number: 70-129544
ISBN: 0-442-25680-9

Manufactured in the United States of America

Published by Van Nostrand Reinhold Company
450 West 33rd Street, New York, N.Y. 10001

Published simultaneously in Canada by Van Nostrand Reinhold Ltd.

15   14   13   12   11   10   9   8   7   6   5   4   3   2   1

Library of Congress Cataloging in Publication Data

Hayat, M   A
    Principles and techniques of electron microscopy.

    Includes Bibliographies.
    1.   Electron microscope—Collected works.   I.   Title.
QH212.E4H38   578′.4′5   70-129544
ISBN 0-442-25680-9   (v. 4)

Preface

This is the fourth volume of a planned treatise on the principles and techniques employed for studying biological specimens with a transmission electron microscope. This treatise departs from the tradition that a book on methodology presents only the contemporary consensus of knowledge. It is written by scholars and when they have anticipated the potential usefulness of a new method they have so stated.

This volume has developed, over the years, through the joint efforts of ten distinguished author-scientists. As a result, a most extensive compilation of methods, developed and used by a large group of competent scientists, has been achieved. The book contains new viewpoints with particular regard to current problems. Areas of disagreement and potential research problems have been pointed out. Discussions of relatively new methods in terms of their application are included. It is within the scope of this volume to provide the reader with detailed methodology for electron probe analysis including possible problems encountered and their remedies.

The basic approach is similar to that in the previous volumes. The methods presented have been tested for their reliability and are the best of those currently available. The instructions for processing the specimens are straightforward and complete, and should enable the worker to prepare his specimens without outside help. It is suggested that entire procedure be read and necessary solutions, media, etc., prepared prior to undertaking the processing. Each chapter is provided with an exhaustive list of references with complete titles, and full author and subject indices are included at the end of the book.

It is encouraging to know that the first three volumes have been favorably received. I feel that this volume will also fulfill its purpose: to provide an understanding of the usefulness, limitations, and potential application of special methods employed for studying the structure, composition, size, number, and location of cellular components and viruses. I hope that it may stimulate a deeper and more refined study of cellular components.

M. A. HAYAT

Contents

3   CORRELATIVE LIGHT AND ELECTRON MICROSCOPY
OF SINGLE CULTURED CELLS
Zane H. Price

4   DENATURATION MAPPING OF DNA
Ross B. Innman and Maria Schnös

5   EXAMINATION OF POLYSOME PROFILES FROM CARDIAC MUSCLE
Kenneth C. Hearn

6   PARTICLE COUNTING OF VIRUSES
Mahlon F. Miller II

7 ULTRAMICROINCINERATION OF THIN-SECTIONED TISSUE
Wayne R. Hohman

8 PREPARATORY METHODS FOR ELECTRON PROBE ANALYSIS
James R. Coleman and A. Raymond Terepka

Contents of

Contributors to this Volume

G. J. Brakenhoff

James R. Coleman

Glen B. Haydon

Kenneth C. Hearn

Wayne R. Hohman

Ross B. Inman

Mahlon F. Miller II

Zane Price

Maria Schnös

A. Raymond Terepka

1. OPTICAL SHADOWING

Glen B. Haydon

Center for Materials Research, Stanford University, Stanford, California

INTRODUCTION

Optical shadowing often provides a dramatic presentation of specimen properties which complement conventional bright-field and dark-field microscopy. The technique requires a simple realignment of the electron microscope and gives a shadowed appearance to the specimen with the impression of three-dimensionality. Although the potential of the technique has not been exploited, the demonstration of several applications (Haydon and Lemons, 1972) has stimulated interest by a variety of electron microscopists.

In the published literature on electron microscopy, a few illustrations of optical shadowing are presented, but only as incidental observations. Hall (1947, 1966) examined the transition region between bright-field and dark-field images of carbon particles in an electron micrograph, and attributed the shadowed appearance to a difference in the amplitude of Fresnel fringes. The same appearance has been illustrated by Dupouy *et al.* (1966, 1968) as conditions met in adjusting the central stop for conical dark-field electron microscopy. Grivet (1972) described the technique as a schlieren method in which the image displays marked shadowing effects termed Foucault contrast, and Andersen (1972) discussed the phenomenon in terms of Fourier theory in which only one half of the Fourier transform is used to create an image. He called the process a single-sideband modulation transfer.

In a theoretical paper, Hanszen (1971) presented the contrast transfer functions for oblique illumination used to achieve optical shadowing. In

1

an earlier paper, he referred to the phenomenon as a single-sideband hologgraphy (Hanszen, 1969); however, his only example is of a light-optical model and the only reference to applications was a possible use in the studies of crystal lattices. Although the range of usefulness may be limited as Hanszen suggests, many types of preparations do reveal additional apparent structural information by optical shadowing.

The technique is not unique to electron microscopy. Indeed, a light microscope can be adjusted in an analogous manner to the electron microscope as described below. Hlinka and Sanders (1970) demonstrated its application in a study of tissue culture cells.

THEORY

Optical shadowing is an application of the schlieren method, in which an opaque edge is placed in the back focal plane of the microscope objective lens and adjusted to allow only a part of the central beam and half of the diffraction pattern to contribute to image formation. The schlieren method is a common technique for visualizing differences in optical paths (for details see an optics textbook: Lipson and Lipson, 1969; Longhurst, 1967). It finds application, for example, in detecting density gradients in an ultracentrifuge cell and refractive index variations as produced by disturbances in a wind tunnel. Another familiar application is the inspection of telescope mirror curvatures using the Foucault knife-edge test. In this test a sharp edge is placed at the focus of the mirror and the image of a point source is viewed from close behind the edge. Imperfections in the mirror surface will appear as shadows on the corresponding portion of the mirror. In a similar way, by placing an opaque edge in the back focal plane of the microscope objective, one is able to visualize variations in the object as a selective shadowing of certain features in the image.

It is possible to gain an understanding of optical shadowing on the basis of relatively simple ray-optics theory. To do so, it is convenient to imagine the irregular surface of a specimen as refracting electrons according to the angle of incidence, as an electron-optical lens effect (Sturkey, 1962; Haydon, 1969). Those areas which refract electrons out of the aperture will appear shadowed while areas which refract into the aperture will appear bright (Haydon and Lemons, 1972).

A more complete understanding of optical shadowing can be gained through the application of wave mechanics. Based on the optical transfer theory of the electron microscope and using this method, Hanszen (1971) developed expressions for both amplitude and phase transfer functions in general and for oblique illumination as is used in optical shadowing. No attempt will be made to present the mathematics of this theory here.

PRACTICE

A practical appreciation of the principles will contribute to the proper adjustment of the microscope for optical shadowing. The relations can be visualized by considering each lens of the microscope separately. Any positive lens can be considered to define two cones determined by its aperture, and the conjugate on-axis image and object points. Aligning of an electron microscope for conventional transmission electron microscopy consists of bringing these cones of all lenses to a common optical axis. In optical shadowing the axis of the cone of illumination is adjusted at a different angle from that of the viewing system.

In a practical example, an appreciation of the cone of illumination and the limiting apertures can be gained in a few minutes using the light microscope. The effect of adjusting the field and condenser apertures and the position of the condenser lens can be visualized easily in a uranium glass cube placed on the microscope stage. The submicroscopic uranium particles in the glass scatter light, thus indicating its path. The condenser lens can be adjusted to image the field aperture in the plane of the specimen, i.e., to bring the crossover spot of the illumination cone to the specimen plane, as is done in bright-field electron microscopy. The field aperture then determines the size of the crossover spot. The angle of the illumination cone is determined by the condenser aperture and is a measure of the numerical aperture of the system. The same relations are present in an electron microscope, but the angle of the cone is very much smaller. Heidenreich (1964, Chapter 1) has a number of illustrations which schematically indicate these relationships.

For optical shadowing, the electron microscope is aligned such that the optical axis of the illumination system produces an in-focus zero-order diffraction spot on the edge of the objective-lens aperture. For the schlieren effect to produce the appearance of optical shadowing, the adjustment is very critical. When the microscope is properly adjusted the image of the specimen is produced by a part of the central beam and the diffraction pattern in only one direction.

In the initial adjustment of the electron microscope it is convenient to start with a holey carbon film as a test specimen, in focus at an intermediate magnification in bright-field (Fig. 1.1A). The conditions for optical shadowing can be observed in the diffraction mode with the specimen in place. First, the objective aperture, which is in the back focal plane of

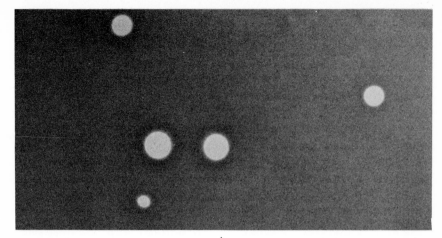

A

B

C

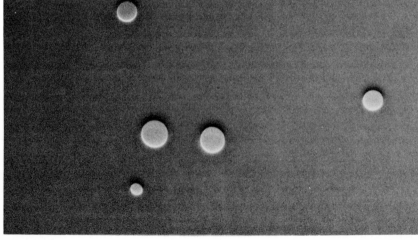

D

the lens, is brought to focus by reducing the current in the intermediate lens. The illumination is then brought to crossover in the same plane as the aperture by adjusting the condenser lens. These are just the conditions used to center the objective aperture in the routine use of many electron microscopes (Fig. 1.1B). The crossover spot can then be moved to the edge of the aperture either by tilting the beam or by moving the aperture (Fig. 1.1C). Upon returning to the imaging mode, optical shadowing is observed in the boundary between bright-field and dark-field images (Fig. 1.1D).

In those electron microscopes in which it is inconvenient to operate in the diffraction mode, it is possible to find the necessary conditions while viewing an in-focus image. Since the image plane of the objective lens is near the aperture, critical tilting of the illumination or movement of the aperture can be accomplished while viewing an in-focus image with the illumination near crossover on the specimen.

With either method, when the crossover point of the illumination is exactly in the plane of the objective aperture, the boundary between the bright-field and dark-field image is broadest. It is in this region that the specimen appears to be shadowed and therefore it is desirable to make this adjustment as carefully as possible.

In those electron microscopes equipped with an electronic beam tilt, bright-field and optically shadowed images can easily be compared by electronically switching conditions. Furthermore, using the electronic beam tilt rather than moving the aperture leaves the imaging lenses optimally aligned with the objective aperture, thereby limiting the effects of spherical aberrations from the objective lens.

As in all electron microscopy, objective astigmatism may be introduced by the objective aperture. Objective astigmatism produced by the electron beam striking the edge is excessive when the older platinum objective apertures are used, because of charging effects. It is probably this difficulty which has given the usefulness of tilted-beam dark-field electron microscopy a bad reputation (Dubochet, 1973). However, with the advent of thin foil apertures clean margins can be found, so that any astigmatism which may be introduced can be corrected even with the beam striking the edge.

The photographic exposure must be adjusted according to the type of

---

Fig. 1.1 (A) A holey carbon film similar to that used as a electron microscope test preparation. (×25,000.) (B) Back focal plane of the objective lens showing a 50-μm aperture centered on the illumination. This is the appearance commonly seen while aligning the aperture with the electron microscope in the diffraction mode of operation. (C) Back focal plane of the objective lens showing aperture with the center of illumination on its edge as it required for optical shadowing. (D) The same holey carbon film is now optically shadowed. (×25,000.)

preparation. If the background is to be bright, the exposure is approximately the same as required for bright-field transmission electron microscopy. However, a greater range of densities and thus better shadow detail can often be recorded by appropriate adjustment of the exposure. Because adjustment of the second condenser is critical in producing the effect of optical shadowing, the optimal exposure can be achieved by adjusting either the gun bias, the exposure time, or a combination of both. Of course, the minimum exposure time of the emulsion used to electrons must be met.

Optical shadowing can be produced at all magnifications and with resolutions up to the potential of the particular instrument in use. With a change in magnification, a readjustment of the microscope alignment may be necessary only if a large change in the objective lens focus is required. The author has had no trouble demonstrating optical shadowing on some half dozen different makes and models of electron microscopes which he has had an opportunity to use. There is no apparent reason that it should not be possible on most other instruments.

INTERPRETATIONS

Care must be taken in the interpretation of the shadowed appearance. The human eye and mind are conditioned to interpret images according to previous experience. In a fascinating book, Gregory (1970) has discussed many of the clues used to give the impression of three dimensions and indicated many of the pitfalls and problems. One of the most common problems with any type of shadowed image is that the first impression of protrusions or depressions is often a matter of the direction from which the micrograph is viewed (Koehler, 1972). This can be easily demonstrated by turning an optically shadowed micrograph upside down. However, with care it is possible to determine the direction of evaporation of electron-dense materials and, using these clues, to interpret protrusions and depressions properly. In optical shadowing it is usually possible to determine the apparent direction of shadowing too. It appears to be from the direction that extends into the dark-field image. This is true if the object has a higher index of refraction than the surrounding medium. In fact, the topographical appearance is dominant because the surrounding vacuum has a lower index of refraction than the specimen. It is possible that the specimen surfaces could be smooth but the specimen might contain regions of differing indices of refraction which would appear shadowed. It is just this property of such local regions of differing indices of refraction to which the German verb *schlieren* refers: whence the name "schlieren method."

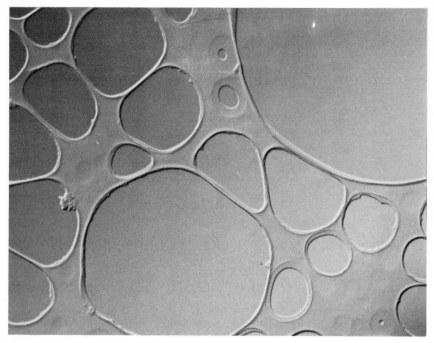

Fig. 1.2 A parlodion net such as is used to support very thin substrates is optically shadowed. The interpretation of holes as holes is easily made with the illustration in this orientation. (×25,000.)

The dramatic three-dimensional appearance can be graphically demonstrated with a parlodion net such as is used to support very thin carbon substrates (Fig. 1.2). For most observers, the first interpretation of holes in the net as being holes is dependent upon the orientation of the print and the anticipated appearance by the observer. Initially illumination is thought to come from the top and cues for the three-dimensional appearance are interpreted on this basis. By turning the illustration upside down and continuing to imagine that the illumination is coming from the top, the holes can be made to appear to be raised plateaus. The micrograph then becomes an optical illusion in which it is possible to "see" either topography. Only with care in the use of other cues can the correct interpretation be made.

A rounded appearance of the edge of a hole in a carbon film fits well with an anticipated structure (Haydon and Lemons, 1972). Unfortunately, the human mind is inclined to jump to the conclusion that a rounded edge has been demonstrated. However, based on theoretical consideration, a specimen with a rectilinear profile could generate an op-

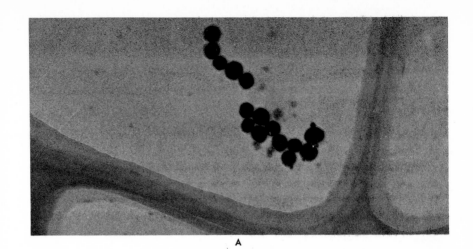

A

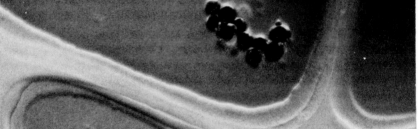

B

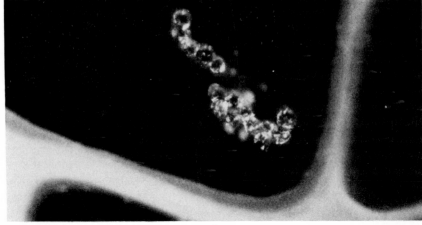

C

tically shadowed appearance similar to that expected from a rounded edge (Goodman, 1968; problem 7-3). This is the condition discussed by Shillaber (1944). A complete theoretical analysis of a real specimen taking into consideration both the amplitude and phase transfer functions of the specimen is beyond the scope of present techniques.

These cautions do not negate the potential value of optical shadowing. Certainly the image is created by properties of the specimen and thus should contribute to an understanding of its structure.

APPLICATIONS

A few applications will serve to illustrate the potential of optical shadowing.

In Fig. 1.3, a preparation of gold colloid which is spread on a thin carbon substrate supported by a parlodion net serves to compare bright-field, dark-field, and optically shadowed microscopy. The appearance of a thin sharp band around the gold particles in bright-field microscopy appears as material piled up around the particles when optically shadowed. In this illustration the details of the parlodion net are most dramatic in the optically shadowed preparation.

A small piece of silica gel powder is illustrated in Fig. 1.4. The two optically shadowed micrographs demonstrate that the apparent direction of the shadow can be changed by moving the central spot to the opposite edge of the objective aperture.

Biological thin sections usually show relatively little in the way of topography. In Fig. 1.5, collagen fibers which are densely stained in the bright-field micrograph appear as individual small protrusions on the surface when optically shadowed. A possible interpretation is that the presence of heavy-metal stains reduces the amount of sublimation of the embedding substrate in the electron beam.

Precipitation of heavy-metal stains on top of thin sections rather than staining of the structures within the section is a problem which is sometimes encountered. An optically shadowed micrograph reveals the problem in a dramatic way (Fig. 1.6).

Finally, a section from the margin of a kidney capillary is compared with conventional bright-field and optical shadowing. The capillary pores appear umblicated and the ribosomes stand up as discrete particles on the surface of the section (Fig. 1.7).

Fig. 1.3 A preparation of gold colloid spread on a thin carbon substrate supported by a parlodion net. (A) Bright-field. (B) Optically shadowed. (C) Dark-field. ($\times$150,000. Preparation courtesy of G. O. Kreutziger and D. A. Taylor.)

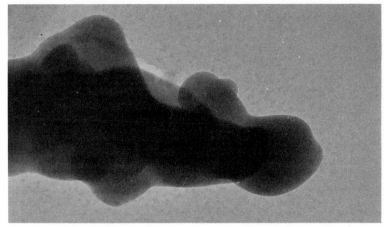

A

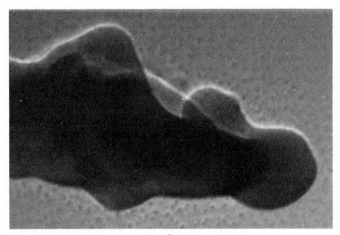

B

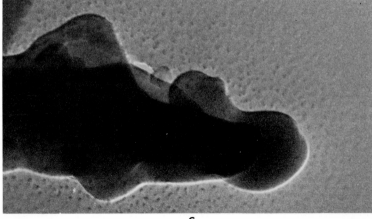

C

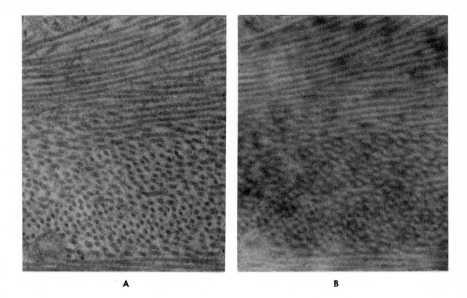

A   B

Fig. 1.5 Collagen fibers from rabbit cornea. (A) Bright-field. (B) Optically shadowed. Although the shadowed appearance is less dramatic in sectioned material, stained areas appear as protrusions on the surface. ($\times$25,000.)

CONCLUSIONS

Optical shadowing provides an additional and dramatic mode of electron microscope operation which may provide more information regarding the specimen. The technique is sufficiently simple that it should be tried on almost all types of specimens. The potential applicability of the technique is extensive rather than limited.

APPENDIX

Outline of Adjustment Procedures for Optical Shadowing

1. Focus on a specimen in bright-field at the desired magnification.

   Then either:

2. bring illumination to crossover on specimen with the second condenser,
3. adjust the size of the illuminated area with the first condenser, and

---

Fig. 1.4 Particle of silica gel powder on a carbon substrate. (A) Bright-field. (B) Optically shadowed from above. (C) Optically shadowed from below. ($\times$150,000.)

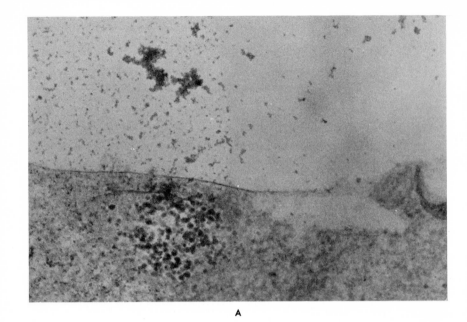

A

B

Fig. 1.6 A tissue section which did not stain properly. (A) Bright-field. (B) Optically shadowed. The optically shadowed image dramatically shows the edge of a drop containing stain aggregates on the surface of the section. (×50,000.)

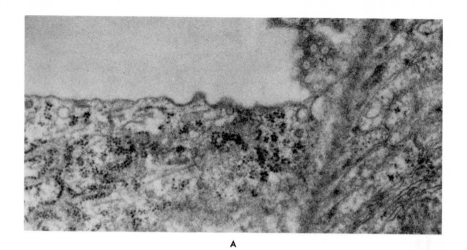

A

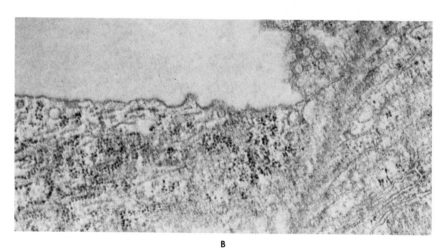

B

Fig. 1.7 Ultrathin sections of a renal capillary. (A) Bright-field. (B) Optically shadowed. (×50,000.) (Haydon and Lemons, 1972)

4. tilt the beam or move the aperture until boundary between bright-field and dark-field appears across the image;

Or

2. image a sharp diffraction pattern of the specimen field,
3. move zero-order diffraction spot to the edge of the aperture with beam tilt or translation of aperture, and

4. return to bright-field.

Finally:

5. Expand optically shadowed boundary with fine adjustment of the second condenser.
6. Adjust objective stigmators and focus.

References

Andersen, W. H. J. (1972). Phase contrast enhancement by single sideband modulation transfer. *Proc. 30th Ann. Mt. Electron Microsc. Soc. Amer.*, p. 616.

Dubochet, J. (1973). High resolution dark-field electron microscopy. *In:* Principles and Techniques of Electron Microscopy: Biological Applications, Vol. 3 (Hayat, M. A., Ed.). Van Nostrand Reinhold Company, New York.

Dupouy, G., Perrier, F., and Verdier, P. (1966). Amelioration due contraste des images d'objets amorphes minces en microscopie électronique: *J. Microscopie*, **5**, 655.

——— (1968). Etude du contraste des objets amorphes en microscopie électronique. *Proc. 4th European Conf. Electron Microsc. Rome*, **I**, 155.

Goodman, J. W. (1968). Introduction to Fourier Optics. McGraw-Hill Book Company, San Francisco.

Gregory, R. L. (1970). The Intelligent Eye. McGraw-Hill Book Company, New York.

Grivet, P. (1972). Electron Optics. 2nd English ed. Pergamon Press, Oxford.

Hall, C. E. (1947). Dark-field electron microscopy. I. Studies of crystalline substances in dark-field. *J. Appl. Phys.*, **19**, 198.

——— (1966). Introduction to Electron Microscopy. 2nd ed. McGraw-Hill Book Company, New York.

Hanszen, K. J. (1969). Einseitenband-Holographie. *Z. Naturforsch.*, **24a**, 1849.

Hanszen, K. J. (1971). The optical transfer theory of the electron microscope: fundamental principles and applications. *In:* Advances in Optical and Electron Microscopy (Cosslett, V. E., and Barer, R., Eds.), Vol. 4, 1–84. Academic Press, New York.

Haydon, G. B. (1969). An electron-optical lens effect as a possible source of contrast in biological preparations. *J. Microscopy*, **90**, 1.

———, and Lemons, R. A. (1972). Optical shadowing in the electron microscope. *J. Microscopy*, **95**, 483.

Heidenreich, R. D. (1964). Fundamentals of Transmission Electron Microscopy. Interscience Publishers, New York.

Hlinka, J., and Sanders, F. K. (1970). Optical shadowcasting of living cells. *Trans. N.Y. Acad. Sci.*, **32**, 675.

Koehler, J. K. (1972). The freeze-etching technique. *In:* Principles and Techniques of Electron Microscopy: Biological Applications, Vol. 2 (Hayat, M. A., Ed.), Van Nostrand Reinhold Company, New York.

Lipson, S. G. and Lipson, H. (1969). Optical Physics. Cambridge University Press, Cambridge.

Longhurst, R. S. (1967). Geometrical and Physical Optics. John Wiley & Sons, New York.

Shillaber, C. P. (1944). Photomicrography. John Wiley & Sons, New York.

Sturkey, L. (1962). Practical considerations in the interpretation of electron diffraction patterns. *Symp. Techniques in Electron Metallog.*, ASTM special technical publ. No. 339.

2. RELATIVE MASS DETERMINATION IN DARKFIELD ELECTRON MICROSCOPY

G. J. Brakenhoff

Laboratory of Electron Microscopy, University of Amsterdam, The Netherlands

INTRODUCTION

The transmission electron microscope (TEM) is used mostly to obtain images at magnifications which are outside the range of the light microscope. The interpretation of these pictures has often a qualitative character. However, under certain conditions, a quantitative approach to the data stored in the electron micrograph is possible. Thus, the addition of quantitative physical or chemical data to the visual information results in unique experimental opportunities.

While for the last decade high-resolution brightfield electron micrographs have been produced as a matter of routine, this is still not generally the case with darkfield. However, the application of darkfield technique to biological studies has recently increased considerably, since various developments have made possible the achievement of high resolution (Thon, 1968; Ottensmeyer, 1969). The theory and application of darkfield has been discussed in detail by Dubochet (1973) in a previous volume of this series. An account of the theory and practice of mass determination from darkfield electron micrographs is presented here.

Mass determination from brightfield electron micrographs has been shown to be possible by Zeitler and Bahr (1962). These authors demon-

strated that, provided certain conditions are met, it is theoretically possible to determine in brightfield the mass of submicroscopic particles falling in the range of 10^{-12}–10^{-18} g. They have presented mass measurements on particles in the range of 10^{-12}–10^{-16} g. At the lower end of the
mass range, they encountered difficulties because of the very low contrast
in the brightfield micrographs. Later in this chapter we will consider their
work in greater detail for purposes of comparison with the darkfield
method presented here.

The darkfield mass determination method (Brakenhoff, 1972) opens
up a new range of particle masses to be investigated, as this method, in
practice, is more sensitive than the brightfield method by a factor of approximately 1,000. Actually, it seems that the lower limits are set by structural inhomogeneities in the films that support the particles to be evaluated.

The theory of the method involves both an investigation of the scattering processes of the electrons in the object and a consideration of the
image formation on the photographic plate by the electron current. In
theory, the electron current can also be measured directly with the aid
of, for instance, a Faraday cage coupled to a sensitive current-measuring
device (Johnson and Parsons, 1969). As is obvious from the later derivation, the darkfield current intensity is (but for the mass thickness limits
set by the specimen), proportional to the specimen mass density per unit
surface. Direct current measurement makes possible the avoidance of
some limitations and difficulties which are inherent in the photographic
plate. However, this method requires specialized measuring apparatus,
and also leads to long periods of electron irradiation of the specimen in
the microscope.

The photographic plate is a universally available medium, and the
mass evaluation from these plates can be obtained away from the TEM.
This can be accomplished either with the simple apparatus to be described later or with more advanced densitometers, some of which now-
adays produce data in a computer-compatible format so as to facilitate
subsequent processing. We therefore considered it practical to base the
present treatment of the darkfield mass determination method upon darkfield images recorded on photographic plate.

DARKFIELD IMAGE FORMATION

Since the theory of the darkfield mass determination method is directly
connected with the scattering processes in the specimen, we will first dis-
cuss briefly the formation of the darkfield image and the methods of
achieving darkfield in the TEM.

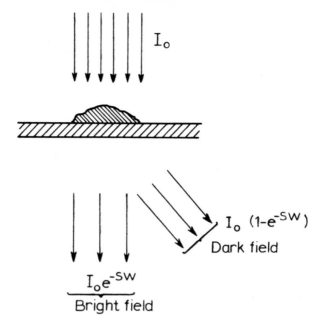

Fig. 2.1 Schematic diagram of the electron scattering process for brightfield and darkfield electron microscopy.

In the TEM the specimen is irradiated by a beam of electrons. The brightfield image is generated by this locally "attenuated" beam due to the removal, by the objective aperture of the electrons scattered by the object. (Fig. 2.1). The darkfield image, on the other hand, is built up by the scattered electrons. In darkfield the intensity of the electron current or the brightness in the image is (except in the case of too high mass thicknesses, which will be discussed later) proportional to the mass thickness.

This proportionality is responsible for the fact that particles with a very low mass density, which are invisible in brightfield, can be imaged in darkfield with satisfactory contrast. This fact accounts for the very high sensitivity of the darkfield mass determination in comparison with the brightfield method. In practice, darkfield can be obtained by various methods, which have been discussed in detail by Dubochet (1973). One of the methods is conical illumination. Figure 2.2 shows this illumination with a centrally located aperture. The mass determination described here is valid for all the darkfield methods. To a first approximation, one can consider the electron scattering in the specimen to be governed by a

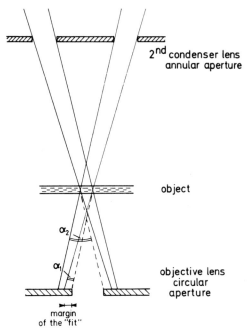

Fig. 2.2 Conical darkfield geometry. An annular aperture is inserted in the second condenser lens. A circular aperture placed in the objective lens intercepts the unscattered part of the beam. It is important to specify the angles α_1 and α_2 between which an electron has to be scattered in order to contribute to the image for a particular darkfield geometry.

single scattering process. The following theoretical treatment is based upon this premise; this assumption will be elaborated upon later.

THEORY

Let E_0 be the current density of the electron beam falling upon an object in the microscope. The mass per unit area w of the object decreases the current density through scattering to the value E_B:

$$E_B = E_0 \exp(-sw) \qquad (2.1)$$

where s ($\mathrm{cm^2/g}$) is the *effective* scattering cross section for scattering outside a certain physical aperture (Hall, 1966). For the electron current scattered outside the aperture, E_D, we can write

$$E_D = E_0 - E_B = E_0 [1 - \exp(-sw)] \qquad (2.2)$$

In darkfield microscopy, the image is formed by a certain fraction of these scattered electrons. If α is this fraction, as determined by the geometry of the darkfield aperture arrangement, we find the following relation for E', the current reaching the image plane from an illumination current density E_0 on the object:

$$E' = \alpha E_D = \alpha E_0 [1 - \exp(-sw)] \tag{2.3}$$

The optical density, D, of the developed grains in the photographic plate is related to the density of the incident electron current through the well known relation (Valentine, 1966)

$$D = b \cdot E' + \delta \tag{2.4}$$

which is valid for density values $D < 1.2$ for most contemporary photographic emulsions at normal conditions. This is true if not too much fogging occurs. The factor b is a constant determined by the photographic material and the processing conditions, while δ takes into account the fogging produced both during development and by the diffusely scattered electrons in the microscope column.

The optical density D of a particular area of the photographic plate is determined from the attenuation of a light beam passing through this area. If I_0 is the original and I the attenuated intensity, then D is given by

$$I_0/I = 10^D \tag{2.5}$$

In practice we will encounter the situation where we have the specimen with the specific mass per unit area w_s supported on a film with specific mass w_f. We can now write for the optical densities D_s and D_f, corresponding, respectively, to the specimen supported by the film and to the film itself, the following relations: combining Eqs. (2.3) and 2.4), and assuming D_s, $D_f < 1.2$, we obtain

$$D_s = b\alpha E_0 \{1 - \exp[-(s_s w_s + s_f w_f)]\} + \delta \tag{2.6}$$

and

$$D_f = b\alpha E_0 [1 - \exp(-s_f w_f)] + \delta \tag{2.7}$$

by subtracting (2.6) from (2.7), we find an extremely simple expression for the relation between the photographic density and the mass density of a specimen present on the supporting film:

$$D_s - D_f = b\alpha E_0 \exp(-s_f w_f) [1 - \exp(-s_s w_s)]$$
$$\approx b\alpha E_0 \exp(-s_f w_f) s_s w_s \tag{2.8}$$

The linear approximation $[1 - \exp(-s_s w_s)] \approx s_s w_s$ is valid for $s_s w_s \ll 1$. Tests to determine up to which mass thickness $s_s w_s \ll 1$ can be considered satisfied we present later.

As we often actually measure the light transmitted through the photographic plate, it is convenient to have an expression which relates the mass to be measured directly to the transmitted light intensities. Using (2.5) and (2.8) we can write for r'

$$r' \equiv (I_f - I_s)/I_f \approx (D_s - D_f) \ln 10 = cw_s \qquad (2.9)$$

The values I_f and I_s correspond to, respectively, the transmission through the supporting film and the transmission through the specimen *plus* the supporting film. $c = b\alpha E_0 \exp(-s_f w_f)\, s_s \ln 10$ is a constant for a certain supporting film thickness and given electron microscopic and photographic conditions. The approximation in Eq. (2.9) is valid if $(D_s - D_f)$ $\ln 10 << 1$. Stated in transmitted light intensities this condition can be written as $I_f / I_s << e$. From the transmission data it can be seen at once whether this condition is satisfied. This latter approximation is mainly one of convenience and not strictly necessary. As discussed in Appendix C, one can, if one has available the transmission data in a form suitable for computer processing, forgo this approximation.

We shall often want to know the total mass W of a submicroscopic particle. This follows from the integral over an area A within which the particle is contained:

$$W = \int_A w\, dA \qquad (2.10)$$

In Appendix A it is shown that if the supporting film thickness is constant or slowly varying, the mass determined from darkfield micrographs can be derived from the relation

$$W = AR'/c \qquad (2.11)$$

with

$$R' = (T_f - T_s)/T_f \qquad (2.12)$$

T_f is the average of the transmission through an area A measured in the immediate surroundings of the specimen, and T_s is the transmission through this area when the specimen is contained in it.

To obtain an absolute mass determination, the value of A/c in (2.11) has to be evaluated with the help of test objects present in the same micrograph. If it is sufficient to know only the weight ratio of the objects 1 and 2, we can write

$$\frac{W_1}{W_2} = \frac{R'_1}{R'_2} \qquad (2.13)$$

It is necessary that the illumination by the electron beam be uniform over the region where mass comparisons are made. If this is not the case, one has to correct for this effect. As shown in Appendix B, we can write for the weight ratio, if the mass thickness of the supporting carbon film can be assumed to be uniform,

$$\frac{W_1}{W_2} = \frac{R_1' \ln T_0/T_{f2}}{R_2' \ln T_0/T_{f1}} \qquad (2.14)$$

with T_0 the light transmitted through an area A of an unexposed part of the photographic film, and T_{f1}, T_{f2} the transmission of the supporting film in the immediate neighborhoods of particles 1 and 2.

EXPERIMENTAL

The main result of the theoretical section above is that the ratio between the values of $R' = (T_f - T_s)/T_f$ for various particles is equal to the mass ratio of these particles (Eq. 12.13). This result is valid if a number of conditions are satisfied. First, single scattering processes of the electrons in the object should be dominant: $sw << 1$; second, the photographic density D of the negative should be lower than 1.2; and last, the nonessential (see Appendix C) condition that $I_f/I_s << e$ ($e = 2.718\ldots$, the base of the natural logarithms). Later we will present two easy tests for the first condition; the last two photographic tests can be performed directly using the micrographs.

In order to determine the value of R' for a particle from the darkfield micrographs, one has to measure T_s, the transmitted light intensity through an area A containing the particle, and T_f, the transmission through the photographic density corresponding to the supporting film. As T_f cannot be measured at the location of the particle itself, one has to estimate T_f from the photographic density in the neighborhood of the particle (see Fig. 2.3).

Taking for I_f the average of the four transmission values T_{f1} to T_{f4}, we get $R' = 1 - 4T_s/(T_1 + T_2 + T_3 + T_4)$. This method is permitted both when the supporting film is reasonably uniform, and when the supporting film thickness has a smooth gradient over the supporting film area (see Appendix A). The fit of the area A need not necessarily be very close around the particle; often it even cannot be, when particles of different shapes are to be compared. The mass ratio between two particles is given by the ratio of their R' values, under the condition that these values be derived from transmission values over equal surface areas A (cf. Eqs. (2.11) and (2.13)).

Depending on the facilities available, one can obtain the T_s and T_f values in various ways. With automatic computer-coupled processing of the darkfield micrographs, one can calculate R' directly after suitable summation of the transmission data points over the area one desires.

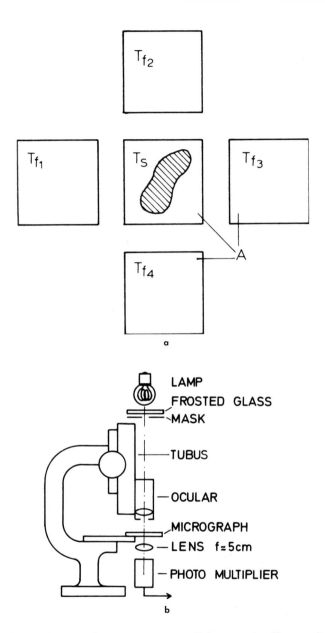

Fig. 2.3a The contribution to the mass measurement of the supporting film over the area A is evaluated by averaging the contributions in the immediate surroundings of the particle.
Fig. 2.3b Schematic diagram of simple transmitted light intensity measuring apparatus.

However, a relatively simple apparatus can be built which operates quite satisfactorily and is sufficient, provided the number of particles to be evaluated is not too large.

The author of this chapter has constructed such a density-measuring device, starting from an old light microscope. A frosted glass plate is uniformly illuminated from the back and is imaged with a suitable demagnification on the micrograph. The lens used for this purpose is a low-power (12×) ocular. This lens is fixed with adhesive tape on the tube of the microscope. The micrograph is framed in a slide frame without glass windows, and the framed micrograph itself is fixed to the microscope table (Fig. 2.3b).

In front of the frosted glass plate we place a mask shaped in such a way that only the desired area A on the micrograph is illuminated. A sharp image of the mask is created on the micrograph by means of the microscope tube height adjustment. The amount of light transmitted through the micrograph is measured with a photomultiplier. The linear response of the photomultiplier over the range used can be checked by noting the photomultiplier output as a function of incident light intensity, when the surface of the mask area is changed in a known manner. The output of the photomultiplier can be monitored on a current meter or recorded on a x-t recorder. With the x-y translation of the microscope table, a desired particle can be brought into the light beam, and the transmission of this particle and the surrounding area can be measured. During these operations, the micrograph is viewed from one side with a 10–20× magnifying binocular.

APPLICATION TO RIBOSOMES

An application of the method is demonstrated on a darkfield micrograph of a specimen prepared from unstained 30S ribosomal subparticles of *E. coli* (Fig. 2.4). The ribosomal particles were sprayed onto an extremely thin carbon film deposited earlier on a freshly cleaved mica surface. The carbon film with the particles was collected on a fenestrated Formvar film, which had been reinforced with carbon. As can be seen, the particles present a variable appearance. Some of them appear as compact entities (e.g., nos. 1 and 6), whereas others appear as loosened structures (e.g., nos. 3–5).

To elucidate further the nature of these particles, relative mass measurements were carried out on several of them. The results are listed in Table 2.1; the light transmission measurements on the micrograph were carried out with the simple apparatus described above. By their relative R' values the particles can be roughly classified into four groups.

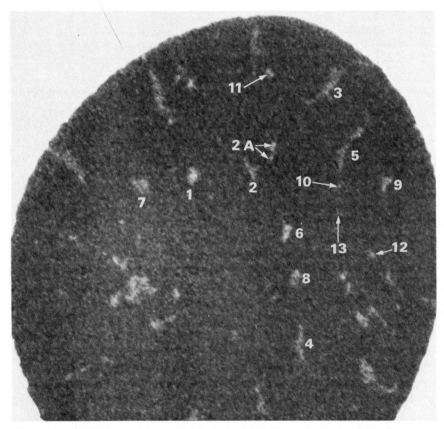

Fig. 2.4 Darkfield electron micrograph of unstained 30S ribosomal subunits of *E. coli*. Particles of variable shape are present. Mass comparisons (see Table 2.2) distinguish between compact particles, unfolded particles, and fragments. ($\times$250,000.)

Group I contains the compact entities 1 and 5, together with the more loosened structures 3,4,5, and 2A and B. Particles 1 and 6 have the typical confirmation of the *E. coli* 30S subunit, which, according to Amelunxen (1971), can occur in an isodiametric or somewhat elongated conformation (see also Brakenhoff, *et al*. 1972). One may, therefore, surmise that particles 1 and 6 represent unfolded versions of the 30S subunit. Particles 3, 4, and 5 are most probably 30S particles disintegrated during handling or deposition on the carbon foil. Groups II–IV contain smaller particles, which are presumably disintegration products of the 30S ribosomal subunit.

If we assume that the group I particles have masses that are approxi-

Table 2.1 Transmission Measurements (Arbitrary Units) and Calculation of R' for the Particles of Fig. 4

Particle	$T_f - T_s$	T_f	R'	Particle	$T_f - T_s$	T_f	R'
	GROUP I				*GROUP II*		
1	13.6	163	0.084	7	6.8	163	0.042
2a + 2b	12.6	168	0.075	8	6.5	165	0.039
3	12.5	173	0.073	9	6	174	0.035
4	11	161	0.069	2a	6	168	0.036
5	10.1	172	0.059				
6	9.6	162	0.059				
	GROUP III				*GROUP IV*		
10	3.8	176	0.022	12	1.6	170	0.0095
11	3.8	173	0.022	13	1.4	175	0.008

Table 2.2 Mass of the Particles Shown in Fig. 4

Mass	I^*	II	III	IV
		Group		
Grams	14×10^{-19}	8×10^{-19}	5×10^{-19}	2×10^{-19}
Daltons	9×10^5	5×10^5	3×10^5	1.2×10^5

* It has been assumed that the particles labelled 1 and 6 of Fig. 4 are 30S ribosomal subunits (see text). The mass given is based on the molecular weight of the 30S subunit as given by Hill et al. (1969).

mately equal to 30S ribosomal subunits, we can assign to each group a certain mass value based upon the value for the molecular weight of the 30S ribosomal subunit given by Hill *et al.* (1969). It is apparent that masses can be determined down to 2×10^{-19} g. The uncertainty in the values of R' is 15 per cent in groups I, II, and III, and 25 per cent in group IV. The biological significance of the grouping of the disintegration products of the 30S particles as presented here will not be discussed; however, it is clear that the information added to the purely visual data is of great value.

DARKFIELD PROCEDURE

The mass determination method described above, is applicable to darkfield micrographs obtained by any technique. The author has always employed the conical illumination geometry as shown in Fig. 2.2, which is preferred for obtaining high-resolution darkfield images. The advantages are as follows: (1) this geometry possesses circular symmetry; (2) the image-forming electrons travel on or near the axis of the microscope (important for low spherical aberration); and (3) in this configuration a comparatively high proportion of the scattered electrons are able to pass the objective aperture. This last property leads to relatively bright image, which is important for reasons of moderate durations of exposure, and focusing. Lack of illumination intensity especially at high magnifications, is often the limiting factor for obtaining a satisfactory focus.

Focusing under marginal conditions has been facilitated by the use of a fiber optic coupled Plumbicon television system in the microscope used for this work (Philips EM 300). The microscope is equipped with a normal filament, which is positioned inside the Wehnelt cylinder of the gun to give a high emission, with some sacrifice of filament life. In any microscope with a double condenser system, conical illumination can be realized by inserting an annular aperture at the location of the second condenser aperture. The unscattered part of the beam should be intercepted with a reasonably close fit (within 1–3 μm) by a circular aperture. Sets of second-condenser annular aperture with a matched circular objective aperture are supplied to us by Micro-Mega Ass., 6 Dommelhoefstraat, Eindhoven, The Netherlands.

The pertinent geometrical data for the micrograph presented in the example on ribosomes are the following: outside diameter annular aperture 1.54 mm, inside diameter 1.24 mm, half-angle of illumination cone 5.15×10^{-3} radian, diameter of the objective aperture 20 μm, margin of fit around the objective aperture 2.5 μm. To pass the objective aperture, electrons had to be scattered over angles between $\alpha_1 = 10^{-3}$ radian and $\alpha_2 = 9 \times 10^{-3}$ radian (see Fig. 2.2). The range $\alpha_1 - \alpha_2$ of the acceptance angles of a darkfield geometry is, from a physical point of view, sufficient to describe such a geometry and should be stated when darkfield micrographs are presented. The microscope used was a Philips EM 300 at 80 kV with the liquid-nitrogen-cooled anticontamination device in operation.

PHOTOGRAPHIC RECORDING PROCEDURES

During the photographic recording of the darkfield image, one should take care that the conditions for the subsequent mass determination are met. While some of them are trivial, others are essential. We will start with the latter.

One should always stay within the linear range of the film (Eq. (2.4)); i.e., $D < 1.2$ with the commonly used types of film. However, with the apparatus used for the photographic density measurement of the mass determination, one can also quite easily measure the response of the film. This is essential when development procedures are employed that deviate from standard, e.g., with a view to influencing contrast or sensitivity.

To measure the response of the film one should expose it to linearly increasing numbers of electrons. This can be accomplished easily by photographing a uniformly lit field with multiple exposures at one exposure time in brightfield, without specimen in the beam. The photographic density should then be linearly proportional to the number of exposures. An example is given in Fig. 2.5 for Kodak FGP roll film developed for 5 min in D 19. The multiple exposures have been realized by screening off an increasing part of the illuminated field from exposures with a movable screen. A good linear behavior is shown up to $D=1$. The absolute density values were determined with the aid of a calibrated step wedge (step size $\Delta D = 0.15$) obtained from UGRA, St. Gallen, Switzerland, which has an absolute accuracy of 0.015.

A few words of caution may be necessary here. The linear response of the photographic plates (within 5 per cent) has been found before for many different emulsion types (Valentine, 1966; Lippert, 1969). The linear dependence does not always pass through the origin. However, this is not important for the method described, since in the calculation of R' the background density (e.g., fogging) drops out. Lippert (1969) has found that the photographic density can depend upon the time that has elapsed between exposure and development. The linear dependence is not affected, but it should be kept in mind as a possible source of inconsistencies between the first and the last exposure on a roll film or a packet of plates. A standard delay between development and exposure of 1 to 2 hr, or keeping the film overnight, should minimize these effects.

If one wants to calculate the mass w directly from the transmission values, i.e., from R', the approximation $(D_s - D_f) \ln 10 \ll 1$, which was derived from Eq. (2.9), should, to a certain extent, be satisfied, depending upon desired accuracy of the results. This has consequences for the photographic densities that can be admitted in the micrograph to be measured. It is essential in the method that the mass ratio of two particles

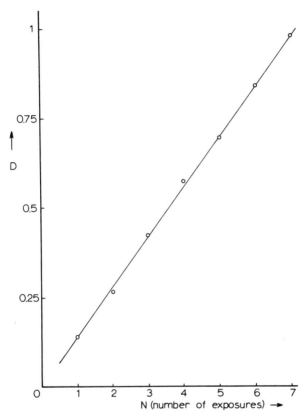

Fig. 2.5 Test of the linear relation between electron current falling upon a micrograph and the resulting photographic density. Film: Kodak FGP; developer: Kodak D 19, for 5 min.

(e.g., 1 and 2) be determined. From the approximation in Eq. (2.9) one can now directly calculate the accuracy with which R'_1/R'_2 is equal to W_1/W_2, depending upon photographic conditions and mass ratios.

The limits for 5% and 10% accuracy for the mass density ratios are plotted in Fig. 2.6. The actual particle mass is not as important as the ratio of the particle masses per unit area. It is apparent that if particles of approximately equal mass density are to be compared, one can admit a large range of photographic densities. If appreciable differences occur, one has to keep the density difference of the reference particle relative to the density of the supporting film (i.e., $D_s - D_f$) within the area set by the boundaries in Fig. 2.6. The linear dependence of the mass ratio upon the R' ratio is valid over a larger range than the linear approximation

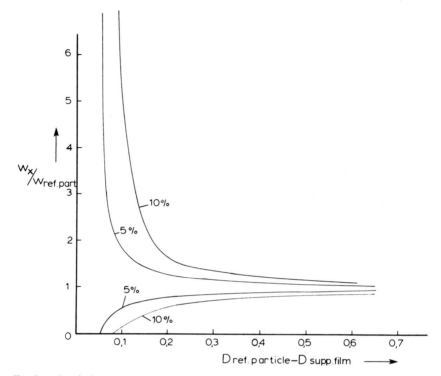

Fig. 2.6 Graph for the determination of the accuracy of the relative mass density measurement between two particles, in relation to the density difference between the reference particle and the supporting film.

$W = AR'/C$ (Eq. (2.11)), since some of the nonlinearity in the latter approximation is canceled in taking the ratio. In general, it is advantageous to take the micrographs in such a way that a comparatively underexposed micrograph results, so that a relatively large range of particle mass densities can be admitted. Then a fine-grained, although slow, film can be used with reasonable exposure times.

Uniformity of development over the negative is important as $r' = cw_s$ (Eq. (2.9)), with c linearly dependent upon the factor b in which the development conditions are also expressed. An easy test is to evaluate micrographs of a uniformly lit field. Some types of densitometer, such as the Optronics or the Digital Film Reader of Joyce Loebl & Co., accept only flexible film and not glass plates, a point which somewhat restricts the choice of photographic film. The magnification should be chosen in such a way that the desired particles are, because of their sizes on the micrograph, easy to evaluate (0.1–1 mm). Possible influences from density fluctuations caused by electron noise (Valentine, 1966) can then generally be neglected.

DETERMINATION OF MAXIMUM PERMISSIBLE MASS THICKNESS

An essential approximation made in the theoretical section is that of the linear dependence of mass density upon scattered electron intensity (Eq. (2.8)). The effective scattering cross section s_s depends on the particular darkfield geometry employed. This is caused by the dependence of the intensity of the scattered electrons on the scattering angle (Lenz, 1954). Therefore, one should know the mass density per unit area w_s up to which the condition $s_s w_s \ll 1$ is satisfied. Two tests are presented below which permit an estimate to be made on the range of validity of the linear approximation in Eq. (2.8).

1. In the case of a fibrillar or tubular biological specimen one can make an unstained preparation and take a micrograph with the darkfield geometry to be tested at a point where the fibers are lying crossed. On the crossover point the mass density will be twice that of the single fiber, and the photographic density difference between the specimen and the supporting film should, if the linearity condition is still satisfied, also be twice as large.

We have used this method with Tobacco Mosaic Virus (TMV). The micrograph is shown in Fig. 2.7 and the measuring results are given in the legend. It is apparent that within the limits of accuracy the linear relation is confirmed. The conclusion can be drawn that at least up to biological mass thicknesses of twice the diameter ($\sim$18 nm) of the TMV rod (i.e., up to $\sim$36 nm of biological specimen) the linear relation is satisfied.

2. To determine the upper limit for the linear dependence between mass density and photographic density, one can use the mass increase due to contamination of a carbon foil. To calibrate the mass increase in terms of equivalent density of a biological specimen, one can again use tubular biological material such as TMV. To illustrate the method, an account of the method employed by us to test a darkfield geometry which permitted passing of electrons scattered in between $\alpha_1 = 1.05 \times 10^{-3}$ and $\alpha_2 = 1.2 \times 10^{-2}$ radian is presented below.

A number of micrographs of an identical area of carbon foil with TMV marker during contamination (anticontamination device was not in operation) were obtained. The photographic density measurements of areas corresponding to the carbon foil D_f and the density difference ΔD_{TMV} of

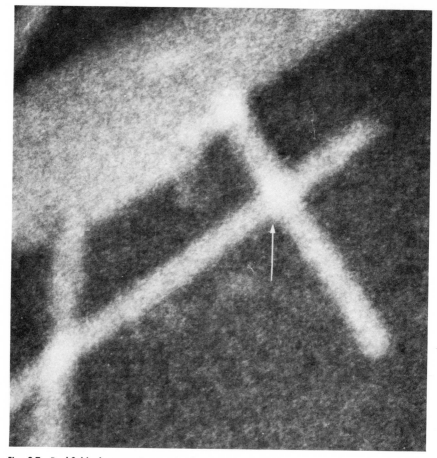

Fig. 2.7 Darkfield electron micrograph of unstained crossed TMV (arrow). The results of the density measurements are as follows:

Density supporting film, D_f: 0.43 $\pm$ 0.02
Density single rod, D_t: 0.41 $\pm$ 0.02
Density crossed rods, D_t: 0.79 $\pm$ 0.02

Darkfield electron microscope data: conical illumination, scattering angles (see Fig. 2.2) $\alpha_1 = 6 - 8 \times 10^{-4}$ radian, $\alpha_2 = 10^{-2}$ radian. ($\times$320,000.)

the TMV rod with the background are plotted as a function of contamination duration in Figs. 2.8a and 2.8b. It is apparent in Fig. 2.8a that the photographic density increases linearly up to contamination durations of ~2 min and then levels off gradually. It is thought for reasons given below that the mass increase due to contamination after $t = 2$ min proceeds

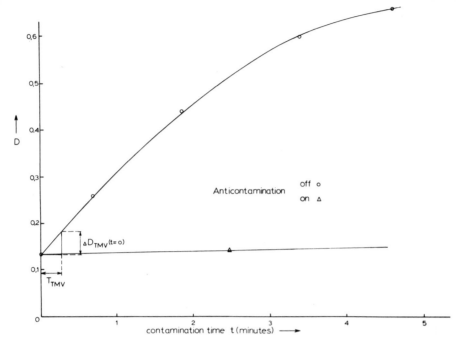

Fig. 2.8a   Increase of the mass thickness of a carbon supporting film with time due to contamination, expressed as the increase in optical density of a series of darkfield micrographs of the contaminating carbon film, taken, under otherwise identical conditions, at the indicated moments.

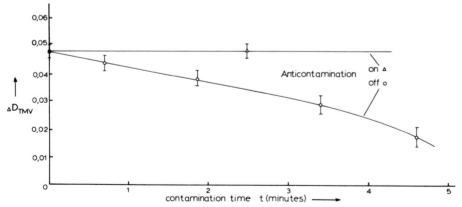

Fig. 2.8b   The dependence of the optical density difference ΔD between TMV on a supporting film and the supporting film itself on contamination. For the derivation of the sw values, see text. If the contamination rate is reduced (Fig. 2.8a) with a nitrogen-cooled anticontamination device, then the mass of the TMV material (Fig. 2.8b) stays constant for a number of minutes.

as before, but that the slackening off of the increase in photographic density is coupled to the breakdown of the linear approximation $1 - e^{-sw}$ $\approx sw$ because the value for the maximum permissible mass thickness is surpassed. Although the actual rate of contamination may differ from day to day depending upon the vacuum history of the electron microscope, we observed a likewise dependence of photographic density on the duration of contamination during such measurement runs. If immediately after such a run the contamination rate of a neighboring fresh field is measured, the contamination rate is the same as present during the linear phase of the measurement run. In view of these results, it can be assumed that the rate of growth of the contamination layer is constant.

With the anticontamination device in operation the contamination is negligible (Fig. 2.8a) and no mass loss of the TMV rod is observed; this can be concluded from the constancy of the ΔD_{TMV} as a function of time (Fig. 2.8b). This is an important observation, as it excludes the possibility that the decrease of ΔD_{TMV} is caused by mass loss under influence of the electron beam. No change within 5% of the intensity of the electron current on the specimen occurred during the measurement series. This was concluded from a comparison of the electron current on the focusing screen from a reference area immediately before and after a measurement run.

To derive a value for $s_s w_s$ from Figs. 2.8a and 2.8b and to determine the maximum mass thickness permitted for a certain degree of accuracy, we start from Eq. (2.8):

$$\Delta D = D_s - D_f \approx b \alpha E_0 \exp\left(-s_f w_f\right) s_s w_s$$

The contamination actually creates a situation where the mass thickness of a specimen present on a "supporting" foil increases steadily with time. Assuming a linear increase in the contamination mass thickness with time, this dependence can be written in terms of units of mass thickness of TMV, w_{TMV} as follows:

$$w_f(t) = w_0 + w_{\mathrm{TMV}} \frac{t}{T_{\mathrm{TMV}}} \tag{2.15}$$

where w_0 is the original mass thickness of the foil, T_{TMV} the time it takes to build up a contamination layer with a mass thickness equivalent to a TMV rod, and w_{TMV} the mass thickness of the TMV rod.

From Eqs. (2.8) and (2.15) we derive, for the time dependence of the density difference ΔD of the TMV rod with the background,

$$\Delta D(t) = b \alpha E_0 \exp\left[-\left(s w_0 + s w_{\mathrm{TMV}} \frac{t}{T_{\mathrm{TMV}}}\right)\right] s w_s \tag{2.16}$$

We assume for simplicity sake that $s_t = s_s = s$, an assumption permitted because both biological and contamination material consist of atoms with low numbers in the periodic table.

Taking the ratio of ΔD at $t = 0$ and $t = t'$ we obtain

$$\frac{\Delta D(t = t')}{\Delta D(t = 0)} = \exp\left(-sw_{TMV}\frac{t'}{T_{TMV}}\right) \quad (2.17)$$

The time T_{TMV} needed for building up a contamination layer equivalent to the mass of a TMV rod can be determined from Fig. 2.8a from the time it takes to add the density $\Delta D(t = 0)$ in the micrograph. We find (Fig. 2.8a) $T_{TMV} = 0.28$ min. Taking now the ratio of $\Delta D(t = 3$ min) ≈ 0.32 and $\Delta D(t = 0) \approx 0.48$ we find from Fig. 2.8b that

$$\frac{\Delta D(t = 3 \text{ min})}{\Delta D(t = 0)} = 0.67 \pm 10\% = \exp\left(-sw_{TMV}\frac{3}{0.28}\right)$$

Calculating the exponent we get $10.7 \times sw_{TMV} = 0.4 \pm 20\%$ and therefore $sw_{TMV} = 0.037 \pm 20\%$.

In Fig. 2.9 the accuracy of the approximation made during the derivation of Eq. (2.8) of $1 - e^{-sw}$ is plotted. For 10% accuracy, for instance,

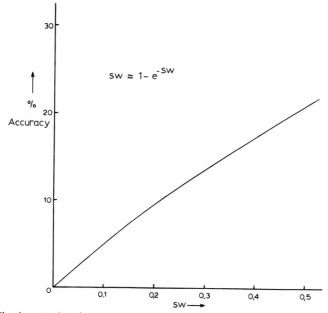

$$sw \approx 1 - e^{-sw}$$

% Accuracy

Fig. 2.9 The deviation from linearity in the approximation of $1 - e^{-sw}$ by sw.

sw values are permitted up to 0.21. For the present test this means that mass thicknesses of biological material up to $0.21/0.037 = 5.5 \pm 20\%$ times the thickness of a TMV rod (18 nm) are permitted for 10% accuracy; that is a thickness of 80–100 nm. Necessary conditions for this method of maximum thickness determination are that the supporting foil before contamination be sufficiently thin and that it still be possible to accomplish the determination of T_{TMV} from $\Delta D(t = 0)$ in the linear part of the scattering intensity dependence on mass thickness.

In order to reduce the relative importance of possible mass loss processes during the interaction of the beam with the specimen, it is advisable to conduct the test on maximum permissible mass under conditions of a relatively high rate of contamination.

The results derived above, at the accuracy specified, are valid for the linear dependence between particle mass and density difference relative to the background film. The ratio of particle masses can, however, be derived from the ratio of the density differences with much greater accuracy; e.g., for particles of equal mass densities the effects of the deviation from linearity cancels out. One should, therefore, consider the accuracy determined from the sw value of a particle with a certain mass thickness as an upper bound. The accuracy limits because of the finite mass density are added to those caused by the determination of the mass ratios from the R' values. In practice, for not too large particles (thickness < 40 nm), the latter effect will be dominant.

SINGLE VS. MULTIPLE SCATTERING

In the theoretical section of this paper it was assumed in the derivation of Eq. (2.9) that only single scattering events take place. In fact, both inelastic and elastic collision processes can scatter the electrons sufficiently to enable them to pass through the darkfield aperture. In particular, the electrons scattered inelastically over small angles will have experienced many scattering events because of the high cross section at these small angles. Several authors (Burge and Smith, 1962; Lenz, 1954) have considered this problem. To a first approximation, as pointed out by Zeitler and Bahr (1957), multiple scattering can be treated in terms of a reduced scattering cross section.

Quantitatively we can say the following: If the value sw is known, one can easily calculate the mean free path for scattering into a particular geometry. In our work, for scattering into the conical darkfield geometry (by accepting electrons scattered in between 1.05×10^{-3} and 1.2×10^{-2} radian at 80 kV), we found that sw expressed in units mass thickness TMV has the value $sw_{TMV} \approx 0.037$. The mean free path l, defined by the condition $swl = 1$, becomes, therefore, in units TMV, $1/sw_{TMV} \approx 28$. In terms of mass thickness of biological specimens, we get $l \approx 28 \times 18$ nm ≈ 490 nm $\pm 20\%$, under the assumption that density of TMV is representative for biological specimens. This means that the assumption of single scattering made during the derivation of Eq. (2.9) seems to be well justified. A quantitative evaluation of the electron scattering in relation to the measured values for the mean free path is given by Brakenhoff *et al.* (1972).

THE INFLUENCE OF DARKFIELD APERTURE GEOMETRY

The maximum permissible thickness of a biological specimen depends upon the effective value of the scattering cross section s_s of the electrons contributing to the image. This cross section depends upon the angle of scattering necessary for the electrons to contribute to the image. This angle is determined by the darkfield apertures. From the differential scattering cross sections published by Lenz (1954), it can be concluded that the upper limit for linear dependence extends up to greater mass thicknesses if the scattering angle of the electrons is increased, while if a smaller scattering angle suffices the limit shifts to smaller mass thicknesses.

For the determination of the range over which the linear dependence is satisfied, the lower limit of the scattering range of the darkfield aperture is the most important one. This is because of the rapidly increasing scattering intensity in smaller scattering angles. To give a practical guideline for 10% accuracy based on the observations with our geometry: mass thicknesses of the biological specimens are permitted with a thickness up to $\sim$ 80–90 nm, if the minimum scattering angle is 10^{-3} radian, the maximum scattering angle not being important. If the minimum scattering angle is larger, a somewhat larger mass thickness can be permitted. For a smaller minimum scattering angle, one should perform one's own check along the guidelines given above.

COMPARISON OF MASS DETERMINATIONS IN BRIGHTFIELD AND DARKFIELD ELECTRON MICROSCOPY

For brightfield electron microscopy, Zeitler and Bahr (1962) have derived an expression comparable to our Eq. (2.12). They found that r, as defined below, is approximately linearly proportional to the mass of the specimen:

$$r \equiv \frac{I_s}{\tau} - \frac{I_f}{\tau} = \frac{I_f}{\tau}\left[\left(\frac{I_f}{\tau}\right)^{\exp(-s_s w_s)-1} - 1\right] \approx a \, s_s w_s \qquad (2.18)$$

where τ is the transmission through a reference area in which only fogging has taken place. The linear approximation in the above equation is valid up to values $s_s w_s \approx 1$. The constant a varies rather strongly with the optical density, except for densities in the range $D = 0.6 \pm 0.2$.

It is advantageous when the area of the plate to be evaluated has a density in this range, although the method is also applicable outside it when a different value is used for the constant a. We see from Table 2.3 that

Table 2.3 Comparison Mass Determination Method in Brightfield and Darkfield Electron Microscopy

	Brightfield*	Darkfield
D	$0.4 - 0.8$	$0 - 1.2$
ΔD†	< 0.4	< 0.25
		or
		$0 - 1.2$‡
$s_s w_s$†	< 1	≤ 0.21
minimum mass (detected)	10^{-16} g	2×10^{-19} g
maximum thickness biological material†	< 4000 Å	< 900 Å§

* Brightfield conditions, Zeitler and Bahr (1962).
† Range permitted for 10% precision.
‡ With computer-assisted film evaluation (see Appendix C).
§ For scatter angles between $\alpha_1 = 1.05 \times 10^{-3}$ radian and $\alpha_2 = 1.2 \times 10^{-2}$ radian.

each method, either in darkfield or brightfield, has its own area of application, and also sets its own demands to be fulfilled for the method to be applicable. The first demand is that in the brightfield method $s_s w_s$ values are admissible up to $s_s w_s \approx 1$ (vs. $s_s w_s \leq 0.21$ for darkfield, 10% accuracy). The reason for this is that in the brightfield method the nonlinear dependence of the scattering intensity on mass thickness is partly compensated by the nonlinearity in the relation between photographic density and transmitted light intensity.

It is clear that the advantages of darkfield electron microscopy are realized chiefly for very small masses. Also, the high sensitivity of the darkfield method renders staining of biological specimens unnecessary. However (see Table 2.3), if the specimen thickness is greater than ~100 nm, the brightfield method is preferred.

A brief comment on the effect of the beam on the specimen is in order. Contamination and loss of mass in the specimen do not affect the result of either brightfield or darkfield methods, provided these effects influence the particles to be evaluated to the same extent. The reason for this is that the methods are essentially comparative.

The brightfield data have been taken from Zeitler and Bahr (1962). They operated in a range of beam voltages between 40 and 100 kV, with apertures such that the half-angles of illumination α were in the 5×10^{-3} to 16.6×10^{-3} radian range. The brightfield data in Table 2.3 are valid at a median value for the scattering cross section of $s_s = 5 \times 10^{-4}$ cm²/g at the voltage and aperture ranges as used by them.

MINIMUM MEASURABLE MASS

On theoretical grounds that a minimum photographic film area is needed because of the inherent granularity of photographic film, together with a certain maximum magnification in the electron microscope, Zeitler and Bahr (1962) arrived at a minimum measurable mass in brightfield electron microscopy of 10^{-18} g under the conditions stated above. The experimental lower limit seems to be 10^{-16} g. If we were to follow Zeitler and Bahr's line of reasoning leading to 10^{-18} g in brightfield, we would arrive at the conclusion that in darkfield microscopy a few atoms having a weight of $\sim 10^{-23}$ g and supported on a perfect amorphous carbon film of 2 nm thickness should actually be measurable.

However, the lower mass limit in darkfield microscopy will, in practice, be considerably higher than this ideal. Factors such as supporting film structure, contamination, and the quality of the darkfield adjustment of apertures play a role. A specific limitation arises as a result of the presence of bright points in high-resolution darkfield microscopy (Thon, 1968; Brakenhoff, 1974), which have sizes in the range of 0.3–0.6 nm.

Therefore, it seems that measurement of particles with a diameter of 1–2 nm on a reasonably thick carbon film (5 nm) should be possible. Such particles would have a mass of $\sim 4.5 \times 10^{-21}$ g (mol. wt. 3,000 dalton) assuming a density of 1 g/cm³ for biological specimens. Possibly the use of single-crystal graphite films (Hashimoto et al., 1971) in which no bright points are visible, offers a way out of the difficulty presented by supporting film structure.

CONCLUDING REMARKS

The method of mass measurement in darkfield electron microscopy is approximately a thousand times more sensitive than the comparable method in brightfield electron microscopy. It seems that the smallest measurable mass is sufficiently low to make this method useful for the measurement of not-too-small macromolecules. The limiting factors are associated with supporting film structure and specimen degradation under the influence of the electron beam. The importance of this degradation effect can sometimes be estimated by comparing micrographs obtained after various durations of irradiation. It is a desirable practice to reduce exposure of the specimen to the beam by, if possible, prefocusing and correction of astigmatism on a nearby field.

Important for many applications is that in darkfield, in contrast to brightfield, no staining of the specimen is necessary. Correction for the mass contribution by the staining reagent is often difficult, if not impossible. The darkfield technique also offers possibilities for studies where the quantitive aspect of the degree of staining is relevant.

The mass determination method presented here is essentially comparative, but, with the help of known calibration particles, it can also yield absolute mass values. The mass measurement of individual particles of macromolecular size and the identification of particles with a variable morphology can yield valuable biological information. The method also permits successful interpretation of fragmentation and aggregation phenomena.

APPENDIX A

Using Eq. (2.9) we can write

$$W = \int_A w_s \, dA = \frac{1}{c} \int_A r' \, dA = \frac{1}{c} \int_A \frac{I_f - I_s}{I_f} \, dA \qquad (2.\text{A1})$$

If I_f is a slowly varying function, we can approximate it by a linear relationship as:

$$I_f = I_{f0} \left(1 + \frac{\epsilon}{2} \frac{\Delta I_f}{I_{f0}} \right) \qquad (2.\text{A2})$$

where I_{f0} is the average intensity over the area to be investigated, ΔI_f is the maximum value of the intensity differences over the area, and ϵ varies from -1 to $+1$ over the area.

We now can write:

$$
\begin{aligned}
W &= \frac{1}{c} \int_A \left\{ -\frac{I_s}{I_{f0}[1 + (\epsilon/2)(\Delta I_f/I_{f0})]} + 1 \right\} dA \\
&\approx \frac{1}{c} \int_A \frac{-I_s[1 - (\epsilon/2)(\Delta I_f/I_{f0})] - I_{f0}}{I_{f0}} \, dA \\
&= \frac{A}{c} \frac{\int_A (-I_s) \, dA + \int_A (\epsilon/2)(\Delta I_f/I_{f0}) \, dA + \int_A I_{f0} \, dA}{A I_{f0}} \qquad (2.\text{A3})
\end{aligned}
$$

The approximation is justified if $\Delta I_f / I_{f0} \ll 1$. I_{f0} is determined by averaging the transmission in areas of the film surrounding the specimen under investigation. If small differences occur between these transmission values, then one can assume that the contribution of the film at the location of the object will have the average value.

The first integral in Eq. (2.A3) is equal to T_s, the total transmission through the area A with the particle contained in it. The third integral is equal to T_f, the transmission corresponding to the supporting film. The contribution of the second integral can be neglected as ϵ varies between -1 and $+1$ over the area, and $\Delta I_f / I_{f0}$ is a small quantity. For the particle weight, we find finally, from Eqs. (2.12) and (2.A3),

$$ W = \frac{1}{c} A R' \tag{2.A4} $$

APPENDIX B

If the illumination by the electron beam over the area in which we wish to compare the particle weights is nonuniform, a correction must be applied. Combining Eqs. (2.9) and (2.8) one obtains

$$ r' = b\alpha \log_{10} E_0 \exp(-s_f w_f)\, s_s w_s \tag{2.B1} $$

Comparing weights determined in areas 1 and 2 with illumination levels E_{01} and E_{02}, and with supporting film weights per area w_{f1} and w_{f2}, one can write for the ratio of the weights contained in these areas, using Eq. (2.11),

$$ \frac{W_1}{W_2} = \frac{E_{02} \exp(-s_f w_{f2})\, R_1'}{E_{01} \exp(-s_f w_{f1})\, R_2'} \equiv B \frac{R_1'}{R_2'} \tag{2.B2} $$

The quantities $E_{01} \exp(-s_f w_{f1})$ and $E_{02} \exp(-s_f w_{f2})$ are, in fact, the intensities in the brightfield image from the respective supporting film areas. If nonuniform illumination is suspected one can take, after a series of darkfield micrographs, a brightfield micrograph under the *same* illumination conditions on the specimen. The brightfield illumination for this micrograph should therefore be effected by replacing the circular aperture in the objective lens by one which is just enough larger that the electrons of the hollow illumination cone can then reach the micrograph. The photographic density on the micrograph is now directly proportional to the quanity $E_0 \exp(-s_f w_f)$, provided that D remains below 1.2. When illumination intensity data are known we can also, after suitable standardization of the method, arrive at mass comparisons of particles on different micrographs.

From a series of darkfield micrographs, taken at constant illumination conditions from various locations on the supporting film, it can sometimes be concluded that the film thickness is uniform while the illumination is nonuniform. Then, as $sw_{f1} = sw_{f2} = sw_f$, we can say that the intensity of the scattered electrons is directly proportional to the illumination intensity, i.e.,

$$E_{01}/E_{02} = \frac{E_{01}[1 - \exp{(-sw_f)}]}{E_{02}[1 - \exp{(-sw_f)}]}$$

As D is directly proportional to the impinging electron current on the photographic plate (see Eq. (2.4), using Eq. (2.5) to tie the density to the transmission values, we find the following relation for B:

$$B = \frac{E_{02} \exp{(-sw_f)}}{E_{01} \exp{(-sw_f)}} = \frac{E_{02}}{E_{01}}$$

$$= \frac{E_{02}[1 - \exp{(1 - sw_f)}]}{E_{01}[1 - \exp{(1 - sw_f)}]} = \frac{\log_{10}{(T_0/T_{f2})}}{\log_{10}{(T_0/T_{f1})}} \qquad (2.B3)$$

where T_0 is the unattenuated intensity of the light beam with area A, and T_{f1} and T_{f2} are the light transmission values through the darkfield micrograph corresponding to the support film in the immediate surroundings of the particles 1 and 2, respectively.

In fact, it is not easy in practice to obtain completely uniform illumination over the darkfield micrograph; 10 per cent differences are hardly visible. The supporting film thickness is, in contrast, nearly always very uniform, provided folds and other irregularities are absent. Therefore, the relation

$$\frac{W_1}{W_2} = \frac{\log_{10}{T_0/T_2}}{\log_{10}{T_0/T_1}} \frac{R_1'}{R_2'} \qquad (2.B4)$$

resulting from (2.B2) and (2.B3) is regularly used in practice.

APPENDIX C

The output of a electromechanical reader of photographic films is usually in the form of a set of light transmission values measured at regularly spaced points covering the micrograph. These data are stored on, for instance, a magnetic tape and become the basic material for subsequent computer processing. An area A which contains the particle to be measured will usually contain a rather large number of these points at which the transmission is measured. From these data one can now calculate the relative mass values of particles over the whole density range of the photographic film below $D < 1.2$, without using the approximation in the derivation of Eq. (2.9).

If a particle with a weight W is situated in an area A we can write

$$W = \sum_{i=1}^{n} w_{si} \frac{A}{n} \qquad (2.C1)$$

with the area A containing n sample points and with w_{si} $(i = 1, \ldots, n)$ the mass density per unit surface of the object at the points i. We now can write Eq. (2.8) in the form $w_{si} = c' (D_{si} - D_f)$, with $c' = 1/[b\alpha E_0 \exp(- s_f w_f)]$. Substituting in (2.C1) we get

$$W = \frac{c'A}{n} \sum_{i=1}^{n} (D_{si} - D_f) = c'A \left[\frac{1}{n} \sum_{i=1}^{n} (D_{si}) - D_f \right] \qquad (2.C2)$$

Rewriting (2.C2) in transmission values with the help of Eq. (2.5) we find

$$W = c'A \left(\frac{1}{n} \log_{10} \prod_{i=1}^{n} \frac{I_0}{I_{si}} - \log_{10} \frac{T_0}{T_f} \right) \qquad (2.C3)$$

where I_{si} is the transmission at point i, I_0 is the light intensity through an unexposed part of the film, T_0 and T_f are the total transmissions through the area A (see also Appendix B), and the notation $\prod_{i=1}^{n}$ indicates the series multiplication

$$\prod_{i=1}^{n} a_i = a_1 a_2 \ldots \ldots \ldots a_{n-1} a_n$$

Taking again (see Eq. (2.B3)) $sw_{f1} = sw_{f2} = sw_f$ we get

$$\frac{c_1}{c_2} = \frac{b\alpha E_{02} \exp(-sw_{f2})}{b\alpha E_{01} \exp(-sw_{f1})} = \frac{\log_{10} T_0/T_2}{\log_{10} T_0/T_1} \qquad (2.C4)$$

and we can write finally for the mass ratio between two particles 1 and 2

$$\frac{W_1}{W_2} = \frac{\log T_0/T_{f2}}{\log T_0/T_{f1}} \frac{\log_{10} T_0/T_{f1} - (1/n) \log_{10} \prod_{i=1}^{n} I_0/I_{1si}}{\log_{10} T_0/T_{f2} - (1/n) \log_{10} \prod_{i=1}^{n} I_0/I_{2si}}$$

The quantities in Eq. (2.C5) can readily be calculated on the computer from the transmission data. Nonuniform illumination over the area of the micrograph is permitted for the use of Eq. (2.C5); in fact, the transmission data of the particles to be evaluated may be derived from different micrographs. The only requirement is that the supporting film of the particles be identical and that no contamination be present on either micrograph (or equal contamination on both).

The author thanks Dr. A. C. v. Dorsten for discussions and Dr. N. Nanninga (University of Amsterdam) for contributing to the biological applications and also for providing (together with J. Pieters, University of Utrecht) the micrograph of the 30S ribosomal subunits.

References

Amelunxen, F. (1971). Untersuchungen zur Struktur der Ribosomen. Ein Beitrag zur Konfirmation der Ribosomen insbesondere der 30S Untereinheit. *Cytobiologie*, 3, 111.
Brakenhoff, G. J., Nanninga, N., and Pieters, J. (1972). Relative mass determination from darkfield electron micrographs with an application to ribosomes. *J. Ultrastruct. Res.*, 41, 238.
Brakenhoff, G. J. (1974). On the sub-nanometer structure visible in high-resolution darkfield electron microscopy. *J. Microscopy*, 100, April.
Burge, R. E., and Smith, G. H. (1962). A new calculation of electron scattering cross sections and a theoretical discussion of image contrast in the electron microscope. *Proc. Phys. Soc.*, 79, 673.
Dubochet, J. (1973). High resolution dark-field electron microscopy. *In:* Principles and Techniques of Electron Microscopy: Biological Applications, Vol. 3 (Mayat, M. A., Ed.). Van Nostrand Reinhold Company, New York and London.
Hall, C. E. (1966). Introduction to Electron Microscopy. McGraw-Hill Book Company, New York.
Hashimoto, H., Kumao, A., and Hino, K. (1971). Images of thorium atoms in transmission electron microscopy. *Japan J. Appl. Phys.*, 10, 1115.
Hill, W. E., Rosetti, G. P., and Van Holde, K. E. (1969). Physical studies of ribosomes from *Escherichia coli*. *J. Mol. Biol.*, 44, 263.
Johnson, H. M., and Parsons, D. F. (1969). Enhanced contrast in electron microscopy of unstained biological material. *J. Microscopy*, 90, 199.
Lenz, F. (1954). Zur Streuung mittelschneller Electronen in kleinste Winkel. *Z. Naturforsch.*, 9a, 185.
Lippert, W. (1969). Erfahrungen mit der photographischen Methode bei der Massendickenbestimmung im Elektronenmikroskop. *Optik*, 29, 372.
Ottensmeyer, F. P. (1969). Macromolecular finestructure by darkfield electron microscopy. *Biophys. J.*, 9, 1144.
Thon, F. (1968). Hochauflösende elektronenmikroskopische Abbildung amorpher Objekte mittels Zweistrahlinterferenzen. *4th Europ. Regional Conf. Electron Microsc.* 1, 127.
Valentine, R. C. (1966). The response of photographic emulsions to electrons. *In:* Advances in Optical and Electron Microscopy, Vol. 1. (Barer, R., and Cosslett, V. E., Eds.). Academic Press. London.
Zeitler, E., and Bahr, G. F. (1957). Contributions to the quantitative interpretation of electron microscope pictures. *Exptl. Cell Res.*, 12, 44.
Zeitler, E., and Bahr, G. F. (1962). A photometric procedure for weight determination of submicroscopic particles. Quantitive electron microscopy. *J. Appl. Phys.*, 33, 847.

3. CORRELATIVE LIGHT AND ELECTRON MICROSCOPY OF SINGLE CULTURED CELLS

Zane H. Price

**Department of Microbiology and Immunology, School of Medicine,
University of California, Los Angeles, California**

INTRODUCTION

The characteristics of the electron beam have severely limited the study of functioning biological material in the electron microscope. Efforts have been made to circumvent these limitations by examining living organisms with high-voltage electron microscopes, but these efforts have not been very successful with bacteria, and they have been less so with cultured animal cells.

The limitations of the study of living cells, *in vitro* inherent in the electron beam can be partially circumvented by resorting to correlative light and electron microscopy. The activity of a selected cell can be followed with the light microscope, and at an appropriate time the cell can be fixed *in situ* and processed for either transmission or scanning electron microscopy. The technique of correlative light and electron microscopy has contributed to a better understanding of the relationship between activities of the living cell and its fixed micromorphology. Correlative microscopy has been used to study Zeiosis (Price, 1967), membrane ruffling (Price, 1968, 1972), and mitosis (Robbins and Gonatas, 1964). Furthermore, the effects of selective alteration on a cell or cell organelles by physical or chemical means can be studied at the level of micromorphology (Bloom, 1960).

Standard monolayer culture techniques are satisfactory for correlative light and electron microscopy of single culture cells *in vitro*. However, special techniques are necessary to assist in identifying the selected cell during the process of dehydration and embedding and in the removal of the resin-embedded cell from the substrate.

THE CULTURE CHAMBER

Correlative light and electron microscopy of single cultured cells *in vitro* begins with the selection of the chamber in which the cells are to be cultured. High-resolution light microscopy can be accomplished only if the cells are grown in a monolayer in small dishes or chambers. The short working distance of the oil immersion objective with a high numerical aperture and its associated condenser necessitates the use of a culture chamber of very shallow dimensions. Thicker chambers or small plastic culture dishes can be used with all objectives, but a condenser with a long working distance must be utilized with these containers. The reduced numerical apertures of condensers of this type means a sacrifice in resolution with oil immersion and most high-dry objectives.

The small perfusion chamber is preferred for correlative microscopy because experimental conditions are more easily controlled during observation, particularly if cell activity is being recorded by time lapse cinephotomicrography. Four chambers of the many that have been designed for cell culture (see Poyton and Branton, 1970, for review), offer the greatest potential for correlative light and electron microscopy of single cultured cells *in vitro*. These are the Rose (Rose, 1954), Sykes–Moore (Sykes and Moore, 1958), Poyton–Branton (Poyton and Branton, 1970), and Dvorak–Stotter (Dvorak and Stotter, 1971) chambers. The Dvorak–Stotter chamber has been specifically designed for use with high-N.A. oil immersion objectives and condensers, either bright- or dark-field. The other three chambers can be used with oil immersion optics, provided thin gaskets or O-rings (1 mm thick or less) are employed. All four of these chambers transmit equivalent images with medium- and low-power objectives.

The ease with which a perfusion chamber can be disassembled is a design feature that is advantageous for correlative microscopy. The substrate must be removed from the chamber and transferred to the secondary marking device without drying of the fixed cells. Another chamber design feature that deserves some consideration is the size and shape of the substrate on which the cells will be grown. Small round cover slips of glass or plastic are easier to manipulate during cell location procedures than are large rectangular ones.

SUBSTRATES

The choice of substrate for the culture of cells for correlative microscopy is dictated by the optical system used for observation, and the resin in which the cell is to be embedded. However, the two groups of resins that are most satisfactory for the embedding of cultured cells for thin-section microtomy have a strong affinity for the two surfaces to which most cultured cells will readily attach. Polyester and epoxy resins are nearly inseparable from glass or styrene. Apparently, separation of the monolayer of embedded cells from a plastic substrate is not always necessary. The plastic substrate can be sectioned along with the embedded cell in some instances. However, the resin–cell layer must be separated from the plastic substrate in the majority of situations.

Plastic substrates have some limitations. Many cell lines do not adhere well to the surface of plastic substrates that are easily released from the embedding resin. It is discouraging to dislodge and lose an experimental cell during processing, simply because the cell was not attached securely to the surface of the substrate. These same plastics also have an internal structure that becomes visible with phase optics.

Most cells adhere well to styrene, and its structure does not interfere with phase optics, but its solubility in acetone and propylene oxide limits its application for correlative microscopy. This limitation can be overcome in most situations by using hydroxypropyl methacrylate (HPMA) as the intermediate solvent between the alcohol series and epoxy resins (Brinkley et al., 1967). Epoxy resins cannot be separated from styrene, but styrene can be trimmed and sectioned as part of the embedment.

Polyester and polypropylene cover slips separate easily from epoxy resins (Firket, 1966), but the crystalline structure of these plastics limits their use with phase optics. The fluoroplastic Aclar 33C (chlorotrifluoroethylene) is probably the most satisfactory plastic for the preparation of coverslip substrates (Masurovsky and Bunge, 1968). This plastic is suitable for phase and fluorescence microscopy, and many cell lines will grow directly on its surface. However, the surface of this plastic must be pretreated with collagen or other material (plasma, celloidin) before some cells will adhere with certainty. The author has replaced the bottom of the lower section of 60-mm styrene Petri dishes with discs of Aclar 33C for the purpose of growing large numbers of cells for selective sampling. The bottom of the styrene Petri dish was removed and replaced with an Aclar disc of appropriate size. The disc was sealed in place with a ring of nontoxic Dow Corning Silastic. Cells grown in these dishes were embedded *in situ* in a thin layer of epoxy, and specific cells were selected and processed as described.

Mica coverslips can be separated from epoxy and polyester resins with ease (Persijn and Scherft, 1965), but the many minute cracks present in every sample of mica that the author has examined make these coverslips unsuitable for phase microscopy.

Glass is the preferred substrate for use in most perfusion chambers, in spite of the fact that polyester and epoxy resins tenaciously adhere to its surface. The strength and clarity of No. 1 and No. 2 cover glasses and the insolubility of glass in the organic solvents routinely used in thin-section microtomy transcend the single disadvantage. The No. 1 cover glass is the most satisfactory substrate for correlative microscopy with incident near-ultraviolet light and fluorescent vital dyes.

Successful separation of cured resin and glass substrate can be accomplished sometimes by sudden cooling of the resin–glass sandwich with dry ice or liquid nitrogen (Howatson and Almeida, 1957), but this procedure is too capricious for use with single-cell techniques. Separation of embedding resin and glass substrate by differential cooling can be facilitated by partially coating the cover glass with Teflon from a spray can (Chang, 1971). Partial, rather than complete, coating of the cover glass with Teflon provides adequate glass surface for cell growth while sufficiently reducing contact of the embedding resin with the glass surface to simplify separation of the two surfaces. This procedure is satisfactory for general sampling of cell monolayers for electron microscopy, but the author has found it unsatisfactory for selected-cell techniques. Cell processes do not always detach from the surface of the glass when the embedding resin is separated from the glass, and the Teflon droplets can interfere with phase microscopy.

A more reliable method of ensuring complete separation of resin and glass without damaging the cells consists of coating the surface of the cover glass with a thin layer of evaporated carbon prior to cell culture (Robbins and Gonatas, 1964). Cells that attach to glass will attach as readily to carbon. The embedding resin adheres to the carbon layer rather than to the glass surface, and the resin–carbon sandwich can be detached from the glass substrate with ease and without damage to the cells. The thin carbon layer can be sectioned along with the embedded cell (Price, 1968). Furthermore, the carbon layer does not interfere with phase optics or radiation.

The layer of carbon is deposited on clean, grease-free cover glasses by vaporizing carbon from carbon electrodes in a vacuum evaporator. The required number of No. 1 or 2 cover glasses is cleaned in 1 : 5 nitric-sulfuric acid solution, washed in distilled water, and dried. The clean cover glasses are positioned in a vacuum evaporator equipped with carbon electrodes and coated with carbon. The thickness of the carbon layer is not critical. A layer sufficiently thick to have an obvious brown coloration is usually adequate.

The carbon layer must be "stabilized" to prevent its premature separation from the glass substrate when wet with culture medium. "Stabilization" and sterilization of the carbon layer are simultaneously achieved by baking the carbon-coated cover glasses at 180°C in a hot-air oven for 24 hr.

DESIGNATION OF THE SELECTED CELL

The method detailed here for the designation of a selected cell for correlative light and electron microscopy was developed for use in conjunction with time-lapse cinephotomicrography of cells cultured in perfusion chambers on carbon-coated cover glasses. The Sykes–Moore chamber (Sykes and Moore, 1959) is used routinely for this purpose in the author's laboratory, but the method is also applicable to the designation of single cells on carbon-coated cover glasses from other types of culture chambers. Minor modifications of the procedure make it equally adaptable for the designation of selected cells growing on plastic substrates.

The Sykes–Moore chamber is sealed with silicon rubber O-rings of 1.5 or 3.0 mm thickness depending on the microscope optics involved, inoculated with a cell suspension, and routinely observed on an inverted microscope. A verticle microscope, equipped with a Ploem system, is used for fluorescence microscopy.

Connection of the chamber to the perfusion system is hampered to some extent by its small size. A plastic holder, designed to fit the opening of the microscope stage, materially assists in surmounting this deficiency. Clear Plexiglass is used for construction of the holder to allow unobstructed observation of the objective during focusing of the inverted microscope. The perfusion chamber is retained in the plastic holder with a small set screw, and the holder is secured to the microscope stage with small clamps (Fig. 3.1). Attachment of the perfusion chamber and holder to the microscope stage is essential to prevent movement of the chamber during engraving of the cell location ring on the underside of the cover glass. A culture chamber that is not securely fastened to the stage is likely to follow the movement of the diamond marker, as it cuts a groove in the cover glass. Attachment of the perfusion chamber to the microscope stage also helps prevent vibration and loss of focus during time-lapse filming.

Experience with this procedure has shown that a cell in a perfusion

Fig. 3.1 Sykes–Moore perfusion chamber and special holder on the stage of an inverted micro-scope. The Plexiglass holder was designed to fit closely into the stage opening. Other stages and other perfusion chambers require a holder of different design. Clear Plexiglass was used for construction of the holder to permit observation of the objective during focusing and the diamond marker during engraving of the cell-localization ring. The chamber holder has a milled-out channel to allow insertion of syringe needles into the chamber. Small clamps and set screws hold the needles in place during their connection and disconnection from the perfusion system to prevent damage to the perfusion chamber O-ring. The chamber holder is secured to the microscope stage with L-clamps. Hamilton fittings and valves connect the perfusion system to the chamber, and control admittance of the fixative from the syringe reservoir.

chamber singled out for study by correlative microscopy is usually diffi-cult to relocate after the cell-bearing cover glass is removed from the chamber. This problem can be avoided by marking the location of the selected cell on the back side of the cover glass, before the perfusion chamber is removed from the microscope. One of the most effective ways of doing this is to use a diamond slide marker to engrave a small ring around the cell (Fig. 3.2). This step in the procedure is not necessary for cells cultured on loose, carbon-coated cover glasses in other types of con-tainers (i.e., screw-capped tubes), unless the cover glasses are taken out of the containers for periodic observation or photography of a selected

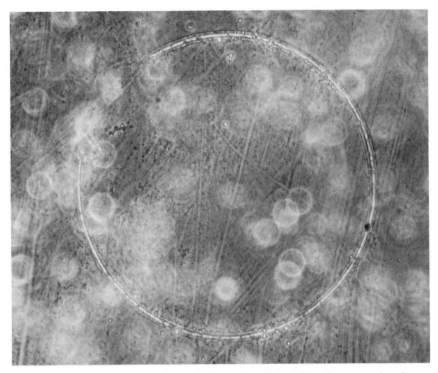

Fig. 3.2 A diamond-engraved circle on the back side of the cell-bearing cover glass from a perfusion chamber. This type of marking serves to designate the location of a selected cell in a closed chamber until the chamber can be opened and the cell relocated and labeled with lines scored in the carbon substrate.

cell. In this case, the cell-bearing cover glass can be placed cell side down on a standard microscope slide, or clamped in the holder that normally is used for making the score marks on the carbon side of the cover glass (Fig. 3.3). Precautions must be taken in engraving a cell-location ring on a cover glass placed in contact with a slide, since there is some risk of cell damage from the pressure exerted by the diamond marker.

Preliminary designation of the selected cell is usually done upon completion of light microscope studies, if only one cell per chamber is being observed. If a number of individual cells are being observed, it is more convenient to mark them at the beginning of the experiment.

Engraving of the ring in the cover glass is facilitated by the use of a centerable quick-change nosepiece. Although this device is not essential to the engraving process, it does aid in circumscribing the cell by centering the marker on the optical axis of the objective. A centerable quick-

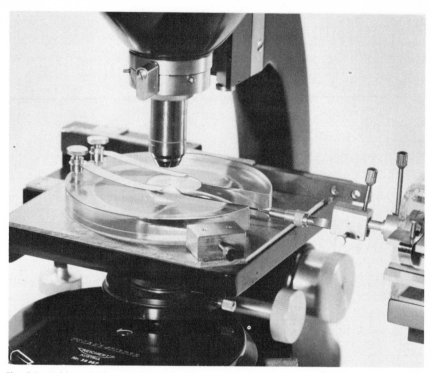

Fig. 3.3 Holder for securing the cover glass during scoring of the cell location marks in the carbon substrate. This holder, like the perfusion chamber holder, is made of Plexiglass to aid visibility. Overall dimensions of the cover glass holder are identical to those of the perfusion chamber holder. The closed rim of the needle channel on which the cover glass rests is coated with silicon rubber. Contact between the cover glass and the silicon rubber ring is maintained by spring stage clips. The cover glass holder is secured to the microscope stage with a $\frac{1}{16} \times 4 \times 4\frac{1}{2}$-inch metal plate and a set screw. The metal plate is attached to the lateral carrier of the stage movement with machine screws. The diameter of the central hole in the metal plate is the same as that of the opening in the stage of the inverted microscope. The micromanipulator for control of the scoring needle need not be as elaborate as the one illustrated. The bent tip of a No. 20 $\times$ 2-inch syringe needle is used to score the carbon layer.

change nosepiece is even more practical if an engraved circle is used instead of score marks to re-mark cell location on the carbon side of the cover glass prior to embedding.

PREPARING THE SELECTED CELL FOR TEM

The cells of the monolayer, including the selected cells, are now ready for fixation, dehydration, and embedding. A cell that is being followed by time-lapse cinephotomicrography is usually fixed *in situ* by perfusion of the culture chamber with phosphate- or cacodylate-buffered 3% glutaral-dehyde. Five milliliters of the fixative is admitted to the culture chamber from a reservoir in the perfusion system (Fig. 3.1) (Price, 1967). If a record of this kind is not part of the experiment, the chamber can be disassembled, and the cell-bearing cover glass immersed in the fixative.

After removing the chamber from its holder, but before disassembly, the general location of the engraved circle on the cover glass should be visually accentuated with a mark from a sharp grease pencil. Secondary marking is not essential but the addition of a more visible dot adjacent to the engraved ring greatly simplifies its relocation during subsequent steps in the procedure.

Fixation is followed by the standard buffer wash. This consists of immersing the cell-bearing cover glass for at least 1 min in each of three changes of slightly hypertonic (350–450 mOsM) phosphate or cacodylate buffer. The cover glass is left in the last change of buffer until ready to indicate the selected cell on the carbon side of the cover glass.

The engraved circle on the back of the cover glass identifies the location of the selected cell only as long as the cell remains attached to the cover glass. The purpose of the engraved circle is destroyed upon removal of the cover glass from the resin embedded cell, so the location of the cell must be re-identified with a mark that will remain with the cell when the embedment is removed from the cover glass. An interruption of any kind in the carbon layer, will accompany the carbon layer when the embedded cells are detached from the cover glass.

Originally the location of the selected cell was re-marked on the carbon side of the cover glass by circling it with an engraved ring similar to the one on the opposite side of the cover glass. However, two engraved rings often proved more than a number one cover glass could withstand. Moreover, the diamond marker frequently pulled cells away from the substrate as it cut through the monolayer. For these reasons, the practice of engraving a circle around the cell on the carbon side of the cover glass was abandoned in favor of scoring four lines in the form of an open cross — | — into the carbon layer (Fig. 3.4). One or two lines is sufficient if the monolayer is not attached securely to the substrate. The lines are scored into the carbon layer with a sharp needle held in a simple micromanipulator. The scoring operation is done with a vertical phase microscope that is equipped with a modified stage and a cover glass holder (Fig. 3.3). The cover glass holder used in the author's laboratory has the same outside dimensions as the perfusion chamber holder, so that the two holders can be interchanged if necessary.

The carbon layer is more conveniently scored if the cover glass is held

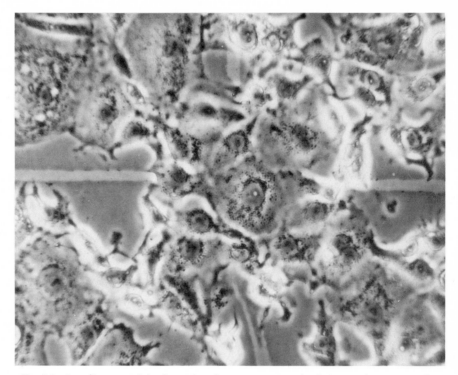

Fig. 3.4 Four lines scored in the carbon layer to permanently designate the position of a selected cell in the monolayer. Fewer lines can be used if the cell or cells are insecurely attached to the carbon layer. The carbon layer and lines are transferred from the coverglass to the embedment along with the cells.

on the microscope stage with the cell side down. The dimensions of the micromanipulator needle and the short working distance of the medium-power objective, make scoring from above difficult, and the needle interferes less with vision if the point approaches the cover glass from below. For these reasons, the cover glass holder was designed to hold the cover glass $\sim \frac{1}{2}$ in. above the stage surface. This permits insertion of the micromanipulator needle from below the cover glass (Fig. 3.3). A phase condenser with a long working distance must be used in combination with this holder. The retention of the cover glass in the holder with the wet side down has an additional advantage. The holder inhibits drying of the cells by acting as a moist chamber.

The wet cover glass is removed from the holder and placed in the buffer upon completion of scoring. The fixed monolayer and the designated cell are osmicated, dehydrated, and infiltrated in the conventional manner, with one difference. Cells intended for TEM are stained with 1 per cent aqueous Alcian Blue 8GX for 30 min following osmication and rinsing. The stain is set by rinsing the cells in 0.01M HCl (Rothman, 1969). An aqueous, non-alcohol-soluble stain is used to prevent leaching of the stain during ethanol dehydration. Staining the cells with a colored dye aids in determining outlines of the embedded cells during final trimming of the block. It has recently been shown that an epoxy-embedded monolayer can be stained with Azure II following polymerization of the resin (Nelson and Flaxman, 1973).

The monolayer of cells is dehydrated in ethanol as rapidly as possible with an exposure of not more than 1 min to each dilution of 70% and 95%. Diffusion of alcohol into the cell is complete during this duration, and longer exposure of the cell to dilute ethanol merely increases the risk of Alcian Blue extraction. The cells can be left longer in absolute ethanol without fear of stain extraction. Although three 1-min changes of absolute ethanol are sufficient for dehydration, the cell-bearing cover glass destined for SEM should be left in absolute ethanol until it can be transferred to a CO_2-critical-point apparatus. The cover glass with a cell destined for thin sectioning and TEM is transferred to propylene oxide or hydroxy-propyl methacrylate (Brinkley et al., 1967), if propylene oxide is contraindicated. Ethanol-saturated cells can be infiltrated with Spurr's epoxy mixture, (Spurr, 1969) without the use of an intermediate solvent.

Infiltration with propylene-oxide–resin mixture is accomplished by immersing the cell-bearing cover glass in a 1 : 1 mixture of solvent and resin, or by inverting the cover glass over a small container filled to overflowing with infiltration mixture. Care must be taken that the volatile solvent does not evaporate completely from the monolayer during transfer to the infiltration mixture.

Infiltration of the cell is completed by placing the inverted cover glass over another container completely filled with resin embedding mixture for 1 hr. Embedding is completed by placing the infiltrated cell in contact with a No. 0 or No. 1 gelatin capsule filled with embedding mixture. The selected cell is positioned reasonably close to the axis of the capsule. This is not difficult to do, since the score marks in the carbon layer are easily found under a low-power dissecting microscope (Fig. 3.5). It is possible to embed a number of selected cells on the same cover glass in this way. The embedding resin is polymerized in a hot-air oven. The duration and temperature of polymerization is governed by the particular resin used for embedding.

The cover glass is removed from the tip of the cured resin block with

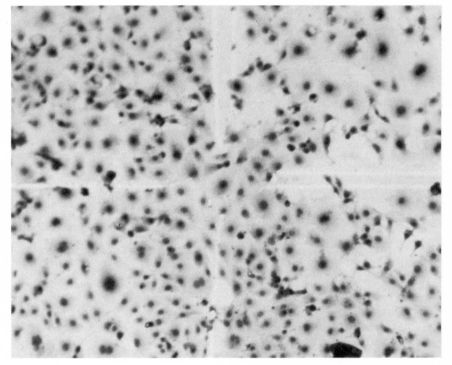

Fig. 3.5 The resin block face with the embedded, blue-stained monolayer as it appears by transillumination of the specimen block. The selected cell is indicated by the lines in the carbon substrate.

gentle finger pressure. It is rare that a cover glass does not come loose with ease. If a cover glass does not loosen with slight pressure, it is usually an indication that the carbon layer contains holes caused by inadequate cleaning of the cover glass. Most of the unwanted cells and excess resin are removed from around the selected cell with a small router. Final trimming of the block to eliminate excess resin is a critical step and must be done with a razor blade under a dissecting microscope. A capsule vise (Fig. 3.6) that permits transmitted illumination of the specimen block makes visualization of the faint, blue-stained cell much easier.

Sections of the cell can be cut in a plane parallel or at right angles to the substrate by the way in which the specimen chuck is mounted in the microtome. Random sections of a selected cell are picked up on grids in the regular way. Serial sections are lifted from the water surface on specimen disks covered with a carbon-stabilized Formvar membrane (Galey and Nilsson, 1966).

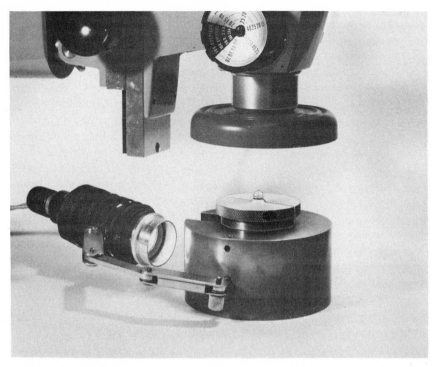

Fig. 3.6 A rotatable vise that permits transillumination of the specimen block. This vise expedites the removal of excess resin and cells from around the cell of interest. The 4-in. diameter by 2-in. thick brass base has a 1-in. hole bored through the center. The 1-in. hole is extended to the exterior of the block by milling out the metal. A 1-in. diameter nylon plug, the top of which is cut at a 45° angle and polished, is inserted into the hole in the base. The polished surface of the nylon plug provides diffuse illumination of the specimen block. A similar plug of Plexiglass with a piece of mirror glued to the angled surface provides nondiffuse illumination. The rotatable clamping units for specimen blocks of various sizes are interchangeable. The source of illumination is a low-voltage microscope lamp.

PREPARING THE SELECTED CELL FOR SEM

A selected cell or cells can be prepared for SEM examination by drying the dehydrated cell with the critical point technique (Anderson, 1951). The author has used CO_2 exclusively for this purpose, but Freon should give equivalent results (Cohen *et al.*, 1968; Cohen, 1974).

The ethanol used for dehydration of the cell is replaced by liquid CO_2 without amyl acetate as an intermediate reagent. Ethanol is completely miscible with liquid CO_2 (Francis, 1954). However, the human olfactory bulb is not as sensitive to ethanol vapor as it is to that of amyl acetate, and it cannot be depended upon to sense the presence or absence of ethanol vapor in the exhaust from the pressure vessel. For this reason, it is more difficult to determine by odor when ethanol has been completely replaced by liquid CO_2. The time elapsed rather than a sense of smell is a more reliable measure of ethanol replacement.

The cell monolayer is kept wet with ethanol during transfer of the cell-bearing cover glass to the critical-point pressure vessel, and until the vessel is filled with liquid CO_2, by the use of a small Y-shaped Teflon holder and a second cover glass (Fig. 3.7). The second cover glass can be cell-bearing or blank. The Teflon Y cover glass holder is 3 mm thick with an overall length of 38 mm. The inside edges of the legs of the Y contain two 0.010-in. slots 0.005 in. apart. The original model of this holder was fitted with a small Teflon pin at the end of one leg to prevent the cover glasses from slipping out of the holder. Experience with the holder has indicated that this pin is superfluous.

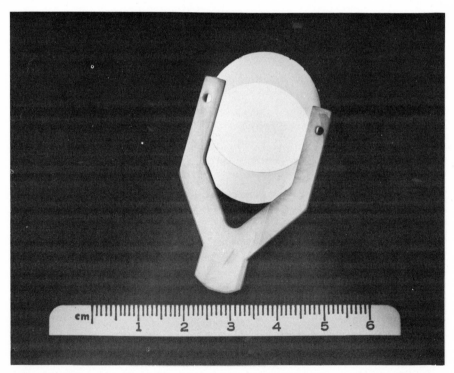

Fig. 3.7 Teflon cover glass holder for drying a monolayer containing a selected cell by the critical-point method. The two cover glasses are retained in 0.010-in. slots spaced 0.005 in. apart. Pins inserted in the small holes in the legs of the holder prevent loss of the cover glasses during processing. The use of these pins was discontinued as unnecessary.

A wet, cell-bearing cover glass is inserted in one of the grooves of the holder with the cell side facing in. A second cover glass is inserted in the second groove of the holder. If the second cover glass has cells attached, it is also inserted with the cell side facing in. The space between the two cover glasses is filled immediately with absolute ethanol from a Pasteur pipette. Capillary action retains the ethanol between the two cover glasses until it is forced out by liquid CO_2. The length of the legs of the Y in relation to the diameter of the cover glasses allows the ethanol to flow out at the top as the liquid CO_2 enters from the bottom.

The cover glass with the dried cells is attached to a SEM specimen stub with double-surface cellulose tape and coated with successive layers of carbon and gold or gold alloy. The carbon layer is not essential for SEM, but it seems to increase the efficiency with which the gold adheres to the surface of the cells, particularly in the deeper convolutions.

The stubs are fastened to a motor-driven turntable which rotates at 120 rpm during evaporation. The turntable is positioned at an angle of $\sim 60°$ to the filament to ensure penetration of evaporated metal into the deeper recesses of the cells. Conduction between the stub and the plated surface of the cover glass is ensured by sealing the edge of the cover glass to the stub with metal glue or silver conducting paint.

CONCLUDING REMARKS

The course of cellular events *in vitro* often can be followed in greater detail if a particular cell in a monolayer can be observed with both light and electron microscopes (Fig. 3.8–3.10). Light microscopy of an *in vitro* cell in a monolayer is essentially routine. The choice of optics and the type of culture chamber is of primary concern. The small separable perfusion chamber has proven most satisfactory for high-resolution light microscopy, and it simplifies preparation of the specimen for electron microscopy.

Transmission electron microscopy requires that the cell observed by light microscopy be fixed, dehydrated, and embedded. This process is also routine, with the exception of labeling the selected cell so that it can be identified during processing and separation of the resin embedded monolayer from the substrate.

Growing the cell monolayer on cover glasses coated with carbon (Robbins and Gonatas, 1964) provides a mutual solution to both of these problems. Lines scored in the carbon substrate to which the cells are attached prior to or after fixation is a convenient way of denoting the location of the selected cell. Four lines scored in the pattern of an open cross

$-\overset{\displaystyle |}{\underset{\displaystyle |}{}}-$ with the selected cell in the center is a label that has proven most

Fig. 3.8 A reconstructed segment of ruffled membrane assembled from enlarged Styrofoam patterns of serial thin sections from a selected cell. The leading edge of the cell is at left.

successful. This type of mark does not weaken the fragile cover glass, and it is easily located at any time during processing. These marks also accompany the selected cell when the embedded monolayer is separated from the cover glass. Furthermore, these marks simplify location of the selected cell in the SEM.

Staining the monolayer with Alcian Blue 8GX (Rothman, 1969) prior to dehydration increases the visibility of the cells in the cured resin, and aids in the removal of surrounding cells and excess resin to isolate the selected cell for thin section microtomy. Post-polymerization staining of the cell monolayer with Azur II is also possible (Nelson and Flaxman, 1973).

Removal of unwanted cells and excess resin to isolate the selected cell for thin-section microtomy is further aided by the use of a specimen block vise that permits transmitted illumination of the block face (Fig. 3.6). Selected cells are examined with the SEM by drying the ethanol-

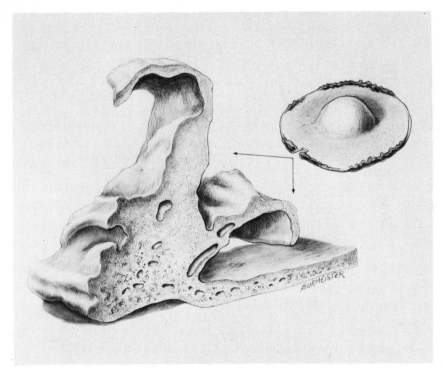

Fig. 3.9   Artists reproduction of ruffled membrane from model in Fig. 3.8.

dehydrated cell with the critical-point methods (Anderson, 1951; Cohen, 1974).

References

Anderson, T. F. (1951). Techniques for the preservation of three-dimensional structure in preparing specimens for electron microscopy. *Trans. N.Y. Acad. Sci.*, Ser. 2, **13**, 130.

Bloom, W. (1960). Preparation of a selected cell for electron microscopy. *J. Biophys. Biochem. Cytol.*, **7**, 191.

Brinkley, B. R., Murphy, P., and Richardson, L. C. (1967). Procedure for embedding *in situ* selected cells cultured *in vitro*. *J. Cell Biol.*, **35**, 279.

Chang, J. P. (1971). A new technique for separation of cover glass substrate from epoxy-embedded specimens for electron microscopy. *J. Ultrastruct. Res.*, **37**, 270.

Cohen, A. L. (1974). Critical point drying. *In:* Principles and Techniques of

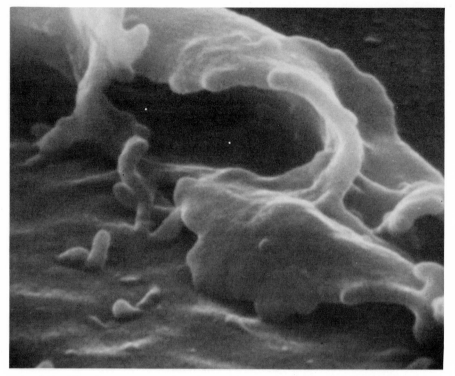

Fig. 3.10   SEM of segment of ruffled membrane from the selected cell. Tilt, 60°. (×29,700.)

Scanning Electron Microscopy, Biological Applications, Vol. 1 (Hayat, M. A., Ed.). Van Nostrand Reinhold Company, New York and London.

————, Marlow, D. P., and Garner, G. E. (1968). A rapid critical point method using fluorocarbons ("Freon") as intermediate and transitional fluids. *J. Microscopie*, 7, 331.

Dvorak, J. A., and Stotter, W. F. (1971). A controlled environment culture system for high resolution light microscopy. *Exptl. Cell Res.*, 68, 144.

Firket, H. (1966). Polyester sheeting (Melinex 0), a tissue-culture support easily separable from epoxy resins after flat-face embedding. *Stain Technol.* 41, 189.

Francis, A. W. (1954). Ternary systems of liquid CO_2. *J. Phys. Chem.*, 58, 1099.

Galey, F. G., and Nilsson, S. E. G. (1966). A new method for transferring sections from the liquid surface of the trough to the supporting film of the grid. *J. Ultrastruct. Res.*, 14, 405.

Howatson, A. F., and Almeida, L. D. (1957). A method for the study of cultured cells by thin sectioning and electron microscopy. *J. Biophys. Biochem. Cytol.*, 4, 115.

Masurovsky, E. B., and Bunge, R. P. (1968). Fluoroplastic coverslips for long term nerve tissue culture. *Stain Technol.*, **43**, 161.

Nelson, B. K., and Flaxman, B. A. (1973). Use of post polymerization whole block staining with Azure II to identify in situ embedded cultured cells for electron microscopy. *J. Microscopy*, **97**, 377.

Persijn, J. P., and Scherft, J. P. (1965). Sheet mica—a nonadherent carrier for surface culture of cells to be embedded in Epon. *Stain Technol.* **40**, 89.

Poyton, R. O., and Branton, D. (1970). A multipurpose microperfusion chamber. *Exptl. Cell Res.*, **60**, 109.

Price, Z. H. (1967). The micromorphology of zeiotic blebs in cultured human epithelial (HEp) cells. *Exptl. Cell Res.*, **48**, 82.

———— (1968). A study of membrane ruffling in cultured cells by time-lapse cinephotomicrography and electron microscopy. *J. Biol. Photog. Assn.*, **36**, 93.

———— (1972). A three-dimensional model of membrane ruffling from transmission and scanning electron microscopy of cultured monkey kidney cells (LLCMK$_2$). *J. Microscopy*, **95**, 493.

Robbins, E., and Gonatas, N. K. (1964). In vitro selection of the mitotic cell for subsequent electron microscopy. *J. Cell Biol.*, **20**, 356.

Rose, G. (1954). A separable and multipurpose tissue culture chamber. *Texas Repts. Biol. Med.*, **12**, 1074.

Rothman, A. H. (1969). Alcian Blue as an electron stain. *Exptl. Cell Res.*, **58**, 177.

Spurr, A. R. (1969). A low viscosity epoxy resin embedding medium for electron microscopy. *J. Ultrastruct. Res.*, **26**, 31.

Sykes, J., and Moore, E. B. (1959). A new chamber for tissue culture. *Soc. Exptl. Biol. Med. Proc.*, **100**, 125.

4. DENATURATION MAPPING OF DNA

Ross B. Inman and Maria Schnös

**Department of Biochemistry and Biophysics Laboratory,
University of Wisconsin, Madison, Wisconsin**

INTRODUCTION

The technique of electron microscopic denaturation mapping depends upon the fact that A-T rich segments of a DNA molecule denature at lower temperatures or pH than G-C rich areas. This fact was first established at the intermolecular level by Marmur and Doty (1962), who showed by optical density melting curves that DNA of high G-C content denatured at a higher temperature than DNA rich in A-T bases. They also demonstrated that there was a linear relationship between denaturation temperature (T_m) and DNA base composition. Should a similar situation exist at the intramolecular level then when an identical set of DNA molecules (in which each molecule has a similar base sequence and base sequence starting point) are partially denatured, each molecule will exhibit denatured sites at similar positions. As the degree of denaturation is increased, regions less rich in A-T should also start to denature, and again these regions should be situated at similar positions from molecule to molecule.

The above hypothesis has been investigated with electron microscopy and found to be valid if certain criteria are satisfied (Inman, 1966, 1967; Inman and Schnös, 1970). When partially denatured molecules are examined in the electron microscope the denatured regions can be seen as two dissociated single strands (Fig. 4.1) and their position can be determined. The denaturation map is simply a linear representation of the molecule showing the position and size of each denatured region (Fig. 4.1).

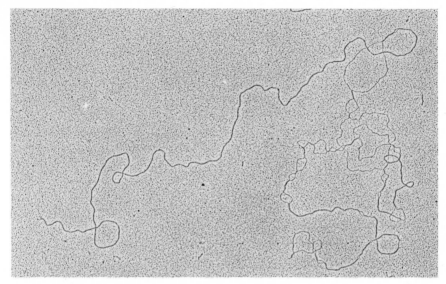

Fig. 4.1   Electron micrograph of partially denatured λ DNA. The denatured sites can be clearly seen as dissociated single-stranded regions which have somewhat less contrast than the double strands. The denaturation map for this molecule is shown below the micrograph (S. Sachs and D. K. Chattoraj, unpublished results). This particular molecule is considerably more denatured than those shown in Fig. 4.3.

Denaturation mapping can be used to answer a variety of questions concerning the structure and function of DNA molecules. This technique yields information regarding base content fluctuations along a molecule. If the heterogeneity of base content is sufficiently large to produce a characteristic denaturation pattern, then a number of further applications are possible. For instance, one can immediately determine from a set of denaturation maps whether or not a particular DNA is circularly permuted. In general, the method offers a useful frame of reference for the study of other events that can be observed on DNA molecules with electron microscopy. The position of replication growing points (Schnös and Inman, 1970, 1971; Dressler *et al.*, 1972; Chattoraj and Inman, 1973), D-loops, heteroduplex loops (Chattoraj and Inman, 1972), repetitive, DNA sequences (Wensink and Brown, 1971; Wake *et al.*, 1972; Skalka *et al.*, 1972), and enzymatic cleavage (Mulder and Delius, 1972) have been successfully studied by denaturation mapping.

DENATURATION MAPPING TECHNIQUES

The method consists of four steps. 1. The molecules must be partially denatured (the degree of denaturation depends upon a number of factors to be discussed later). The partially denatured DNA must be stabilized by the presence of formaldehyde and/or formamide. 2. The sample must be prepared for electron microscopy by a modification of the protein film technique (Kleinschmidt *et al.*, 1962) which will insure that the single stranded regions can be visualized. 3. Partially denatured molecules are photographed in an electron microscope. 4. Electron micrographs are measured and denaturation maps constructed. The details of each of these steps will now be discussed.

Partial Denaturation

Partial denaturation by heat, high pH, or formamide have so far been investigated. We find that the high-pH method leads to more reproducible results than thermal or formamide denaturation. Thermal denaturation seems to give rise to a higher background of random denatured sites, while formamide tends to produce denaturation at the ends of some molecules (the latter effect has not been investigated in detail). The following conditions have been used in our laboratory for partial denaturation of a variety of bacteriophage DNA's.

1. *Thermal Denaturation* (Inman, 1966)

A solution containing 0.1 M NaCl, 0.0067 M KH$_2$PO$_4$, 0.0034 M EDTA, 10% HCHO (Matheson, Coleman, and Bell), and DNA (O.D. 260 nm = 0.01–0.005) at a final pH between 6.5 and 7.5 is heated in a constant-temperature bath for 10 min. The sample is quickly cooled in an ice bath, and then prepared for electron microscopy (see below).

The degree of denaturation will be determined by the temperature and duration of heating; due to the presence of formaldehyde the DNA will not reach an equilibrium state in 10 min and, therefore, the degree of denaturation will be time dependent. For a heating period of 10 min, DNA with an average G-C content of 50% will begin to denature at 48°C. The process will be almost complete at 59°C. The ionic strength can presumably be varied somewhat, but it should be remembered that the width of the helix-coil transition decreases as the salt concentration increases (Marmur and Doty, 1962). This may also be reflected in the denaturation maps as a decrease in the differential melting of A-T and G-C rich regions.

2. *High-pH Denaturation* (Inman and Schnös, 1970)

A solution containing 0.0678 M Na$_2$CO$_3$, 0.0107 M EDTA, and 33.9% HCHO (Matheson, Coleman, and Bell) is adjusted from pH 9.9 to the required denaturation pH with 5 M NaOH. This solution is then mixed with DNA (in 0.02 M NaCl, 0.005 M EDTA) to yield O.D. 260 nm = 0.03–0.01. The final salt concentration should be 0.02 M Na$_2$CO$_3$, 0.0067 M EDTA and 10% HCHO. This mixture is then left at 23°C for 10 min and cooled in an ice bath before preparation for electron microscopy (see below).

As indicated earlier, because of the presence of formaldehyde, the denaturation is time-dependent. For DNA's with 50% G-C content a pH of 11.0 produces the first sign of denatured sites, whereas at pH 11.4 almost complete denaturation results. Again, higher salt concentrations might decrease the differential melting of A-T and G-C base pairs.

The sample described above can be neutralized prior to preparation for electron microscopy, but we have obtained somewhat better results when the neutralization is omitted and the solution is spread at a high pH.

Preparation of Specimens

A number of modifications of the protein film technique (Kleinschmidt *et al.*, 1962) have been recommended for visualization of single-stranded regions in DNA. The procedure used for studies involving heteroduplex molecules (Davidson and Szybalski, 1971; Davis *et al.*, 1971) should be adequate for examination of partially denatured DNA, and the particular modification (Inman and Schnös, 1970) used in our laboratory is described below. It has been our experience that all methods, including the one described below, suffer from unknown effects which can produce drying artifacts from time to time.

1. *Preparation of Carbon-Coated Mica Disks*

Mica disks ($\frac{3}{8}$ in. diameter) are punched out from freshly cleaved mica sheets (Ladd Research Ind., or Ted Pella Company) with a blanking punch held in a drill press. Prior to punching, the mica is placed, freshly cleaved side down, on a piece of paper and taped to it along two edges. The paper protects the mica during and after the punching operation. The mica disks are then cleaned with a stream of air and placed in a vacuum, and a carbon film is evaporated onto the freshly cleaved side. The thickness of the carbon film must be determined by trial and error; it should be as thin as possible, but sufficiently thick to survive the various manipulations which follow.

A $\frac{1}{16}$ in. outer annulus of the carbon film is scratched away from each disk. This step produces a hydrophilic ring around the carbon film which is needed when the DNA is later picked up. The scratching operation can be easily performed by a rotating annular brush and vacuum hold-down device for the disk. Large numbers of these carbon-coated disks can be prepared at one time and then stored at ~10% humidity. Our impression is that these disks work best after they have "aged" for a week or more; however, after approximately 4 months they may no longer yield satisfactory results.

2. Sample Preparation

The partially denatured DNA solution (described above) is mixed with an equal volume of formamide (Matheson, Coleman, and Bell) and cytochrome C (2 × cryst. and lyophilized, Cal. Biochem.) is added to a final concentration of 0.01%. Formamide greatly enhances the visualization of single-stranded DNA (Westmoreland et al., 1969). A 1.2 ml drop of double-distilled water is placed on an indentation on a Teflon block. The Teflon block (15 cm × 15 cm × 0.2 cm) is bonded to an aluminum support which in turn is mounted on leveling screws. Twenty-four hemispherical indentations are milled into the top surface of the Teflon, each being 1.9 cm in diameter and ~0.1 cm deep. Before use, the Teflon block is cleaned by brushing the surface with detergent and rinsed thoroughly with distilled water.

A 0.005 ml aliquot of the spreading solution is transferred by a 0.1 cm i.d. capillary pipet to a clean glass rod (0.3 cm diam. and drawn to a fine but rounded tip at one end), as indicated in Fig. 4.2. The glass rod should be kept wet to assure that the spreading solution wets the rod

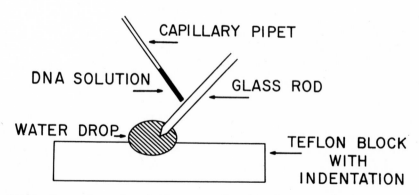

Fig. 4.2 Diagram showing how a DNA-containing film is formed on a water drop.

evenly. The rod is carefully removed from the side of the drop. The surface of the film is compressed by removing 0.1 ml of water from within the drop, using a syringe with a fine needle. The DNA samples can then be picked up by touching a carbon film (evaporated onto a mica surface as described previously) to the surface of the drop. The mica disk with carbon film and drop of liquid is immersed and washed in ethyl alcohol and dried in a stream of warm, dry nitrogen gas. Alternatively, the mica can be dipped into alcohol and dried on a filter paper.

The DNA sample, still mounted on the carbon-coated mica disk, is rotary-shadowed with 3 cm pure platinum wire (0.008 in. diam.) wound around a tungsten filament (0.035 in. diam.). The distance from the filament to the center of the rotating table is 10 cm and the height from the table to the filament is 1 cm. After shadowing, the carbon film (containing the cytochrome film and the DNA) is floated off the mica on a clean water surface and picked up on a specimen grid, and is now ready to be examined in the electron microscope.

3. *Electron Microscopy*

Platinum-shadowed samples can be examined with normal bright-field electron microscopy at magnifications between 4000× and 7000×. The exact magnification is determined by the length of the molecules, the type of camera, and the method used to measure the micrographs. The preparative method described above yields single-stranded regions which can be clearly seen but which have lower contrast than double-stranded DNA. One should guard against basing conclusions on measurements made from molecules which appear to have been aligned during the preparative procedure. In such molecules, the observed length may not be meaningful and the single-stranded denatured regions may not be well resolved.

Measurement, Computation, and Display of Denaturation Maps

Denaturation map data can be obtained from the electron micrographs in a variety of ways. One of the simplest methods consists of projecting the micrographs onto a paper screen, tracing the molecules with a pencil, and finally measuring the marked positions of the denatured sites with a map measure. The map measure should have a swiveling handle which acts as a caster to facilitate the measurement (Keuffel and Esser Co., stock item 620300). The denaturation maps can then be constructed by hand or the data can be read into a computer for plotting. Several of the more sophisticated desk-top calculators can be interfaced to digitizing devices which generate x and y coordinates as a hand-held pen is moved over a projected image of the micrograph. Three such systems have been tested in our laboratory:

1. A Hewlett Packard 9820 (or 9810) calculator and 9864A digitizer can be used to digitize and manipulate data from electron micrographs. A Hewlett Packard 9862A plotter can also be interfaced to this system to allow the immediate plot output of denaturation maps. The system has the disadvantage that the digitizer is opaque and, therefore, the micrograph cannot be rear-projected onto the working surface of the digitizer.

2. A Hewlett Packard 9820 (or 9810) calculator can be interfaced to a Numonics Corporation Graphics Calculator. This system has the advantage that the Numeonics digitizer can work on any surface and rear projection is therefore possible (similarly the digitizer could be used directly over a television screen from an image-intensifier–electron-microscope system). Again, a Hewlett Packard 9862A plotter would allow the immediate plot-out of denaturation maps once the digitizer has been traced across the molecule and the site positions recorded.

3. The Numonics Corporation Graphics Calculator can be obtained with an option that allows lengths to be calculated by the Digitizer rather than by an additional desk-top calculator. These lengths could then be output to a computer via a teletype or the data could be treated manually.

APPLICATIONS

Denaturation maps of a number of DNA's have been published. Examples are DNA from bacteriophages λ (Inman, 1966, 1967; Inman and Schnös, 1970), P2 (Inman and Bertani, 1969; Schnös and Inman, 1971), 186 (Chattoraj and Inman, 1973), human papilloma virus (Follet and Crawford, 1967), adenovirus type 2 (Doerfler and Kleinschmidt, 1970), polyoma virus (Follet and Crawford, 1968; Bourquignon, 1968), SV40 (Mulder and Delius, Yoshiike et al., 1972), T7 (Dressler et al., 1972) and ribosomal DNA of Xenopus laevis (Wensink and Brown, 1971). The degree of reproducibility of the denaturation maps observed with these different DNAs is quite variable and presumably depends upon the magnitude of base content fluctuation along each type of molecule. Figure 4.3a shows maps of λ, P2, 186, and P4 phage DNA. One can see at once that these molecules are not circularly permuted. The histogram in Fig. 4.3b is a convenient way to illustrate the average position of denatured sites in a large set of molecules (Inman, 1967). Detailed maps at a variety of degrees of denaturation have been obtained for λ (Inman, 1967; Inman and

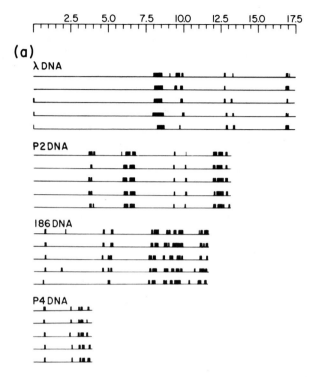

Fig. 4.3a Examples of phage DNA denaturation maps. Each horizontal line represents a DNA molecule and the black rectangles show the size and position of denatured sites. The maps have been normalized to the average length of the particular native DNA. Denaturation conditions were pH 11.28 for 10 min, pH 11.11 for 10 min, pH 11.17 for 20 min, and pH 11.09 for 10 min for λ, P2, 186, and P4 phage DNA, respectively.

Schnös, 1970), P2 (Inman and Bertain, 1969), 186 (Chattoraj and Inman, 1973) and adenovirus 2 (Doerfler and Kleinschmidt, 1970); in these cases information about relative base content heterogeneity is, therefore, available.

The difference in helical stability between thymine- and 5-bromouracil-containing λ DNA can be demonstrated by partial denaturation mapping (Inman and Schnös, 1970). This may be a useful tool for the location of 5-bromouracil segments in DNA.

As already mentioned the most interesting application of the technique is the use of denaturation mapping as a frame of reference for the study of biological phenomena. During replication phage DNAs are frequently circular. If such molecules are partially denatured, the point on the circle corresponding to the mature ends can be located. Figure 4.4 shows a partially denatured λ DNA circle. When this circle is measured, starting

(b)

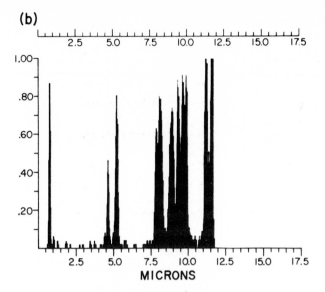

Fig. 4.3b Histogram average of a large set of partially denatured 186 DNA molecules (pH 11.17 for 20 min).

from the point marked by the arrow, the resulting map (Fig. 4.4) agrees well with the maps obtained from mature linear molecules (Figs. 4.1 and 4.3a). Thus, the position marked by the arrow (Fig. 4.4) corresponds to the original ends of this molecule.

Figure 4.5 shows a λ DNA molecule which has been isolated during its first round of replication. Again, after partial denaturation, the map allows one to pinpoint the position of the original ends. The denaturation map thus provides a frame of reference to study the sites of DNA synthesis present at the two branch points (or growing points) shown in Fig. 4.5. After investigating a number of such replicating λ DNA molecules, it was possible to deduce that replication started from a unique position and usually proceeded bidirectionally around the circle (Schnös and Inman, 1970). Similarly, replication starts from a unique position in both P2 and 186 phage DNAs. In these cases, however, the structure is a single-branched circle and replication is always unidirectional (Schnös and Inman, 1971; Chattoraj and Inman, 1972). Figure 4.6 shows an example of a replicating 186 DNA molecule. The positions of the mature ends and the origin of replication are indicated by the arrows in the micrograph and denaturation map, respectively. Denaturation mapping has also been used to locate the origin of replication T7 DNA (Dressler *et al.*,

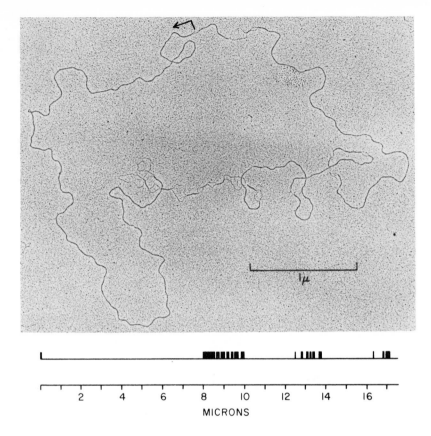

Fig. 4.4 Electron micrograph of a partially denatured circular λ DNA molecule (pH 11.05 for 10 min at 23°C). The denaturation map was measured from the point marked by the arrow. The tail of the arrow also points to the position of the mature ends of the molecule (cf. the linear λ maps shown in Fig. 4.3 or in Fig. 4.1).

1972). In this case, the structure is linear and replication proceeds bidi-rectionally.

When λ phage infects *E. coli* under conditions where replication is blocked by mutation in either the phage or host, circular molecules with D-loops can be observed (Fig. 4.7), and again the position of this single-stranded loop can be determined with reference to the denaturation map (Inman and Schnös, 1973).

The position of deletion, insertion and addition mutations can be de-termined by the study of heteroduplex molecules (Davidson and Szybal-ski, 1971; Davis *et al.*, 1971; and Westmoreland *et al.*, 1969). A partially

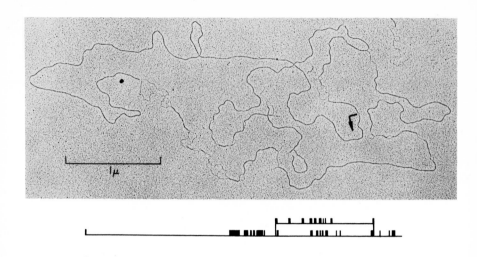

Fig. 4.5 Electron micrograph of a partially denatured replicating λ DNA molecule (pH 11.05 for 10 min). The tail of the arrow marks the position of the mature ends and the denaturation map provides a frame of reference to locate the position of the two replicating growing points.

denatured P2/P2 *vir 22* heteroduplex is shown in Fig. 4.8 (Chattoraj and Inman, 1972) and the relationship between the position of the *vir 22* deletion (arrow) and the denatured map is shown in Fig. 4.9. In addition, it is often possible to directly demonstrate the position of a deletion by noting the absence of a particular part of a denaturation map. Figure 4.10 shows P2 *del 1* denaturation maps and by comparison with the normal P2 map (Fig. 4.3) one can deduce that the extreme right end of P2 is missing in P2 *del 1*. In fact, heteroduplex experiments indicate that the *del 1* deletion does not run completely to the end of the molecule, but stops 0.3% short of the right end (Chattoraj and Inman, 1972).

Direct proof that certain types of DNA contain repeated sequences can be obtained by denaturation mapping. For instance, the ribosomal DNA from *Xenopus laevis* consists of repeated segments (Wensink and Brown, 1971). It also has been shown that under certain conditions DNA can be isolated from infected cells in a form where the molecules have more than one unit length (Wake *et al.*, 1972; Skalka *et al.*, 1972).

In addition to the examples given above, there are a number of further applications which involve identification of sheared phage DNA fragments (Geisselsoder *et al.*, 1973), identification of the position of enzy-

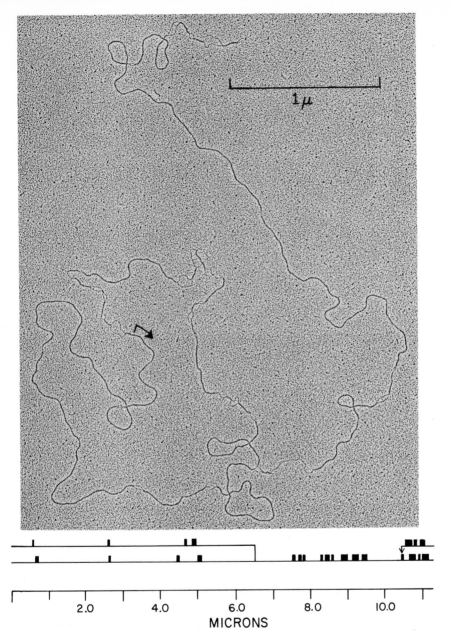

Fig. 4.6 Electron micrograph of a partially denatured replicating 186 phage DNA molecule. The denaturation map has been measured from the point marked by the arrow. The map provides a frame of reference for the study of the position of the replication growing point. The origin of replication is shown by the arrow in the denaturation map. The single-stranded region situated on the tail segment close to the growing point is not due to denaturation but results from the replication process.

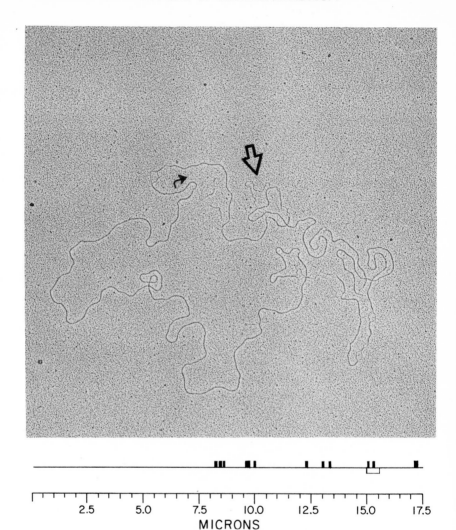

Fig. 4.7 Electron micrograph of a partially denatured λ DNA circle containing a D-loop. These molecules are often found when λ phage infects E. coli with a mutation in one of the genes required for replication (in this case dna B). The D-loop is marked by the open arrow. Its position can be deduced from the denaturation map which was measured from a point corresponding to the mature ends of λDNA (closed arrow).

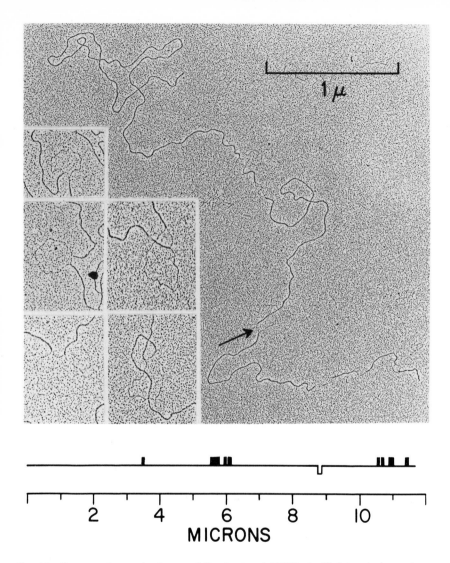

Fig. 4.8 Electron micrograph of a partially denatured P2/P2 *vir* 22 heteroduplex molecule (pH 11.12 for 10 min at 23°C). The heteroduplex loop is marked by the arrow. Its position with respect to the denaturation map is shown by the open rectangular well. The inserts show further heteroduplex loops. Figure 4.9 shows more examples of maps of partially denatured heteroduplex molecules.

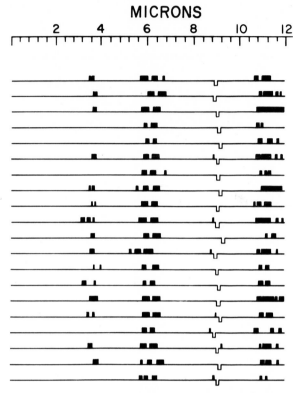

Fig. 4.9 Denaturation maps of P2/P2 *vir* 22 heteroduplex molecules. The black rectangles show the position of denatured sites while the open rectangular wells represent the position of the single-stranded loops caused by the *vir* 22 deletion (pH 11.12 for 10 min at 23°C).

matic cleavage (Mulder and Delius, 1972) and partial denaturation of double-stranded RNA (Shapshak and Inman, unpublished observation).

References

Bourguignon, M. (1968). A denaturation map of polyoma virus DNA. *Biochim. Biophys. Acta,* **166,** 242.

Chattoraj, D. K., and Inman, R. B. (1972). Position of two deletion mutations on the physical map of bacteriophage P2. *J. Mol. Biol.,* **66,** 423.

——— (1973). Origin and direction of replication of 186 bacteriophage DNA. *Proc. Nat. Acad. Sci. U.S.A.,* **70,** 1768.

Davidson, N., and Szybalski, W. (1971). Physical and chemical characteristics of lambda DNA. *In:* The Bacteriophage Lambda (Hershey, A. D., Ed.), p. 45. Cold Spring Harbor Laboratory.

MICRONS

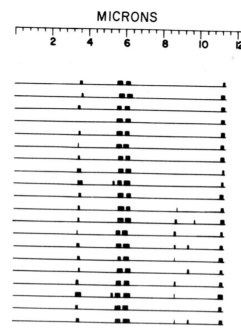

Fig. 4.10 Denaturation maps of P2 del 1 DNA. P2 del 1 contains a deleted DNA segment at the right end of the molecule, and this is reflected in their denaturation maps (compare for instance the right ends of the above maps with those shown for P2 in Fig. 4.3a or Fig. 4.9).

Davis, R. W., Simon, M., and Davidson N. (1971). Electron microscope hetero-duplex methods for mapping regions of base sequence homology in nucleic acids. *In:* Methods in Enzymology, Vol. 21 (Grossman, L., and Moldave, K., Eds.), p. 413. Academic Press, New York.

Doerfler, W., and Kleinschmidt, A. K. (1970). Denaturation pattern of the DNA of adenovirus type 2 as determined by electron microscopy. *J. Mol. Biol.*, **50**, 579.

Dressler, D., Wolfson, J., and Magazin, M. (1972). T7 DNA replication. *Proc. Nat. Acad. Sci. U.S.A.*, **69**, 998.

Follet, E. A. C., and Crawford, L. V. (1967). Electron microscope study of the denaturation of human papilloma virus DNA. II. The specific location of denatured regions. *J. Mol. Biol.*, **28**, 461.

―――― (1968). Electron microscope study of the denaturation of polyoma virus DNA. *J. Mol. Biol.*, **34**, 565.

Geisselsoder, J., Mandel, M., Calender, R., and Chattoraj, D. K. (1973). *In vivo* transportation patterns of temperate phage P2. *J. Mol. Biol.*, **77**, 405.

Inman, R. B. (1966). A denaturation map of the λ phage DNA molecule determined by electron microscopy. *J. Mol. Biol.*, **18**, 464.

———— (1967). Denaturation maps of the left and right sides of the λ DNA molecule determined by electron microscopy. *J. Mol. Biol.*, **28**, 103.

————, and Bertani, G. (1969). Heat denaturation of P2 bacteriophage DNA; compositional heterogeneity. *J. Mol. Biol.*, **44**, 533.

————, and Schnös, M. (1970). Partial denaturation of thymine- and 5-bromouracil-containing λ DNA in alkali. *J. Mol. Biol.*, **49**, 93.

————, and Schnös, M. (1973). D-loops in intracellular λ DNA. *In:* DNA Synthesis *In Vitro* (Wells, R. D., and Inman, R. B., Eds.). University Park Press, Baltimore.

Kleinschmidt, A. K., Lang, D., Jacherts, D., and Zahn, R. K. (1962). Preparation and length measurements of the total deoxyribonucleic acid content of T₂ bacteriophages. *Biochim. Biophys. Acta*, **61**, 857.

Marmur, J., and Doty, P. (1962). Determination of the base composition of deoxyribonucleic acid from its thermal denaturation temperature. *J. Mol. Biol.*, **5**, 109.

Mulder, C., and Delius, H. (1972). Specificity of the break produced by restricting endonuclease R₁ in simian virus 40 DNA, as revealed by partial denaturation mapping. *Proc. Nat. Acad. Sci. U.S.A.*, **69**, 3215.

Schnös, M., and Inman, R. B. (1970). Position of branch points in replicating λ DNA. *J. Mol. Biol.*, **51**, 61.

———— (1971). Starting point and direction of replication in P2 DNA. *J. Mol. Biol.*, **55**, 31.

Skalka, A., Poonian, M., and Bartl, P. (1972). Concatemers in DNA replication; electron microscopic studies of denatured intracellular λ DNA. *J. Mol. Biol.*, **64**, 541.

Wake, R. G., Kaiser, A. D., and Inman, R. B. (1972). Isolation and structure of phage λ head-mutant DNA. *J. Mol. Biol.*, **64**, 519.

Wensink, P. C., and Brown, D. D. (1971). Denaturation map of the ribosomal DNA of *Xenopus laevis*. *J. Mol. Biol.*, **60**, 235.

Westmoreland, B. C., Szybalski, W., and Ris, H. (1969). Mapping of deletions and substitutions in heteroduplex DNA molecules of bacteriophage λ by electron microscopy. *Science*, **163**, 1343.

Yoshiike, K., Furuno, A., and Suzuki, K. (1972). Denaturation maps of complete and defective simian virus 40 DNA molecules. *J. Mol. Biol.*, **70**, 415.

5. EXAMINATION OF POLYSOME PROFILES FROM CARDIAC MUSCLE

Kenneth C. Hearn

Cardiothoracic Institute, University of London, England

INTRODUCTION

In the myocardium, as in other issues, ribosomes exist singly or in aggregates of varying number. In recent years, interest has been shown in the degree of aggregation in relation to the rate of protein synthesis in a variety of physiological and pathological conditions. Current methods of examining ribosomal profiles require separation of aggregates of varying sizes on sucrose density gradients and measurement of these fractions in terms of their ultraviolet absorption (Noll, 1969). Such methods generally produce a series of rather ill defined peaks (Earl and Morgan, 1968; Moroz, 1967; Schreiber *et al.*, 1968). An example of such a profile is shown in Fig. 5.1. A quantitative comparison of different profiles which relies on planimetry is at best somewhat inaccurate because the separation of different fractions is incomplete even with refined methods such as zonal centrifugation (Norman, 1970).

A method has therefore been developed for describing ribosomal profiles which is essentially quantitative and will allow a more reliable comparison of profiles from cardiac muscle in different experimental situations. The method is simple, although somewhat time-consuming, and involves the isolation of ribosomes, their preparation for electron microscopy, and the performance of a differential count on the ribosomal population from the electron micrograph.

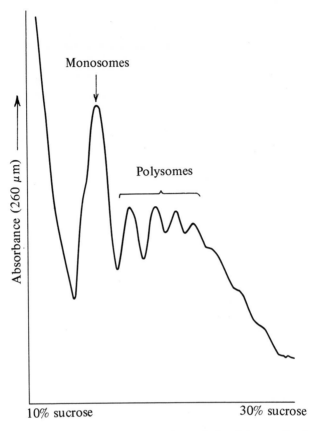

Fig. 5.1   Ribosomal profile from rat heart muscle as displayed by centrifugal analysis on a linear sucrose gradient.

METHODS

Isolation of Ribosomes

The use of isolated ribosomes to analyze the regulation of protein synthesis in animal cells was pioneered by Korner (1958). His methods have been adapted by Earl and Morgan (1968) to make them more suitable for the study of protein synthesis in cardiac muscle. It is essentially the technique of Earl and Morgan which has been employed for the isolation of ribosomes.

Male Sprague–Dawley rats weighing between 180–200 g are killed by cervical fracture. The hearts are quickly dissected and perfused with 5 ml Krebs–Henseleit Solution at 4°C. Perfusion is performed by cannulating the aorta with a 1 mm diameter tube attached to a 5 ml syringe from which the fluid has been gently expelled. All subsequent homogenization procedures are carried out at 4°C. Having removed the atria, great vessels, and any adherent fat, the ventricles are weighed and then finely chopped with scissors and washed with homogenizing medium containing 0.2 M sucrose; 0.02 M Tris/HCl, pH 8.2; 0.1 M KCl; 0.05 M NaCl; 6 mM Mg($CH_3COO)_2$; 1 mM EDTA. A 25% homogenate is prepared using an Ultra-Turrex homogenizer at a setting of 80 V for two periods of 10 sec separated by a pause to avoid heating.

The homogenate is centrifuged at a low speed (10,000g for 10 min in an MSE 50 Super Speed centrifuge) to remove unfractured cells, nuclei, mitochondria, myofibrils, and membranes. The supernatant is carefully removed and made 0.2 M with respect to NH_4Cl and 0.3% with respect to deoxycholate. The sample is then layered over 1 M sucrose (containing 0.02 M Tris/HCl, pH 8.2; 0.1 M KCl; 0.05 M NaCl; 6 mM Mg($CH_3 COO)_2$; 1 mM EDTA) and the ribosomal particles are collected by high-speed centrifugation (105,000g for 90 min). The top layer is carefully aspirated, followed by the 1 M sucrose layer, leaving a pale yellow transparent pellet. The sides of the tube and the ribosomal pellet are washed with homogenizing medium and the washings aspirated as before. When required for microscopy the pellet is redispersed by agitation in ice-cold resuspending medium (0.02 M Tris/HCl. pH, 7.6; 0.1 M KCl; 0.04 M NaCl; 6 mM Mg ($CH_3COO)_2$; 1 mM EDTA; 0.25 M NH_4Cl; and 6 mM 2-mercaptoethanol) in a ratio of 1 ml of buffer for each initial 1 g of cardiac tissue.

Electron Microscopy

A 50 μm grid was used for all experiments. Initially, the specimen support film was of Formvar-backed carbon. More recently, however, an unsupported carbon film has been used in order to improve the contrast and resolution.

One drop of the ribosomal suspension is layered onto the grid, and after 60 sec this is stained with freshly prepared 1% aqueous uranyl acetate. The stain is allowed to remain in contact with the specimen for 90 sec, after which the grids are given three separate washes in distilled water each lasting 10 sec. The grids are then allowed to dry on filter paper, and when dried, are ready for microscopy.

To ensure the random selection of micrographs each field on the grid is selected under low magnification ($\times$ 20,000). The magnification is then increased to ×63,000 and the area photographed. A print of each micrograph enlarged twice is used for counting.

The Differential Count

Counting the numbers of ribosomes and polysomes in each print is helped by placing a Perspex sheet with 4 cm squares etched on one side over the print. The particles within each square and those touching the left and top of the square are counted. Any particles touching the remaining sides are included in adjacent squares. The results of the individual counts of monosomes and polysomes of different sizes are calculated as percentage of the total number of ribosomes and polysomes counted in each print.

RESULTS

Figure 5.2 shows an electron micrograph prepared for counting in the above manner. In the following discussion we have used the term "monosome" to describe a single isolated ribosomal unit. The term "polysome" is used to describe an assembly of such units. Such an assembly may be further defined as a diribosome, triribosome, tetraribosome, etc., according to the number of units in it. The term "ribosomal body" is used to describe both monosomes and polysomes.

Three separate studies were carried out to determine the randomness and degree of variation in the distribution of the ribosomal population prepared in this fashion. The results are summarized in Table 5.1. In the

Table 5.1   Summary of the Distributions of the Ribosomal Population*

Ribosome group	*Count (%)†*		
	Intrasquare count	*Intragrid count*	*Intergrid count*
Monosomes	42.1	45.6	42.5
	(4.8)	(7.3)	(5.1)
Diribosomes	22.1	22.2	22.9
	(3.4)	(4.3)	(2.4)
Pentaribosomes	5.1	4.4	5.3
	(2.6)	(2.3)	(2.0)
Decaribosomes	1.1	0.4	1.2
	(1.0)	(0.7)	(1.1)

* All polysome groups were counted, but only three representative sizes are shown here.
† Standard deviations in parentheses.

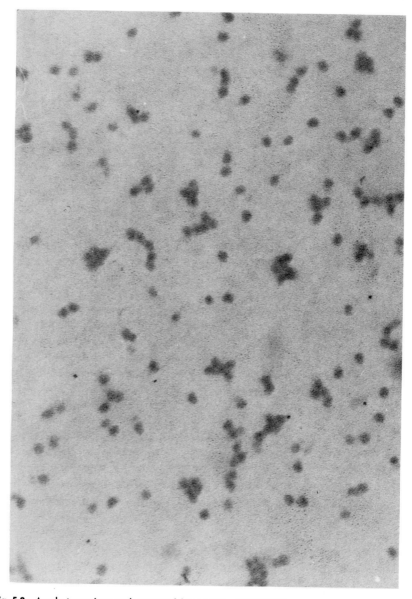

Fig. 5.2   An electron micrograph prepared for counting. ($\times$126,000).

first study ten electron micrographs were taken at regular intervals across an individual square of the grid starting at the left-hand edge and ending at the right-hand edge. Differential ribosomal counts were made on each micrograph. The results showed that the distribution of ribosomes across the square was random. The mean percentage of monosomes was 42.1, with a standard deviation of 4.8.

In the second study a series of 30 electron micrographs were taken diametrically across the entire grid. Each micrograph was taken at the center of an individual square. The results showed that the distribution across the grid was random, the mean percentage of monosomes being 45.6, with a standard deviation of 7.3.

The third study was designed to indicate the variation between grids. Five grids were spread with the same ribosomal preparation. Four micrographs were taken from each grid at points representing the corners of an imaginary square located at the center of the grid, and each side of which was composed of ten small grid squares. The mean percentage of monosomes in the five grids was 42.5, with a standard deviation of 5.1 between the grids.

The method has been used to examine ribosomal profiles from a number of different tissues; Fig. 5.3 contrasts the profiles from rat heart and rabbit heart. For simplicity, polysomes containing six or more ribosomes have been counted in two groups rather than individually. The first group consists of polysomes containing between six and ten ribosomes, and the second contains polysomes with eleven or more ribosomes. In the rabbit heart the percentage of monosomes and diribosomes is higher than in the rat heart, while the percentage of all the larger polysomes is lower. This difference is most marked in the polysomes with six or more ribosomes.

CONCLUDING REMARKS

The polysome is a fragile complex of ribosomes linked by messenger RNA which could easily be disrupted by harsh homogenization conditions and by ribonucleases released from the parent tissue. However, the methods of isolation that are currently practiced reduce the possibility of degradation to a minimum (Earl and Morgan, 1968; Arlinghaus and Ascione, 1972).

The ionic environment during isolation is also critical. Magnesium ions are essential for the assembly of the polysome, but an excess of magnesium produces a nonspecific aggregation of ribosomes which appear superficially similar to polysomes (Takanami, 1960; Huston et al., 1970). Using the differential counting technique the effect of magnesium ions

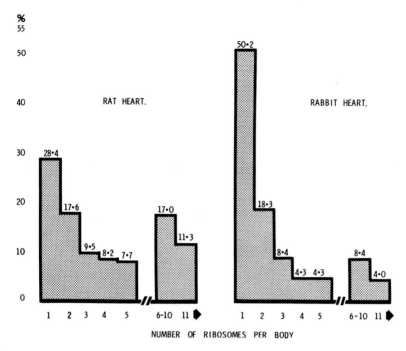

Fig. 5.3 Comparison of ribosomal profiles from rat heart and rabbit heart as displayed by the differential counting method.

in the homogenization and resuspension media (Hearn *et al.*, 1972) has been examined and a concentration of 6 mM Mg^{2+} has been found to be optimal.

Another possible source of error arises during the deposition of the ribosomal suspension on the carbon support film (Horne, 1965). Aggregation of particles is produced when their concentration in the suspension is too high. It is easy, however, to dilute the suspension and prepare grids further when this occurs.

The positive staining technique for ribosomes is essentially that described by Burghouts *et al.*, (1970). It has been found to be superior to more complex staining procedures, the chief advantages being its speed and simplicity.

The results of the counting show the feasibility of the method. There is a random distribution of ribosomes between grids, within grids, and within squares, and the counts are reasonably reproducible. It would seem that this method of numerical analysis of ribosomal preparation would have a number of applications in the study of protein synthesis in the myocardium when its quantitative nature will be more valuable than the qualitative results afforded by centrifugal separation in sucrose density gradients.

References

Arlinghaus, R. B., and Ascione, R. (1972). Tissue culture polyribosomal systems. *In:* Protein Biosynthesis in Nonbacterial Systems (Last, J. A., and Laskin, A. I., Eds.), p. 29. Marcel Dekker Inc., New York.

Burghouts, J. T. M., Stols, A. L. H., and Bloemendal, H. (1970). Free and membrane-bound polyribosomes in normal and Ranscher-virus-infected mouse spleen cells. *Biochem. J.*, 119, 749.

Earl, D. C. N., and Morgan, H. E. (1968). An improved preparation of ribosomes and polysomes from cardiac muscle. *Arch. Biochem. Biophys.*, 128, 460.

Hearn, K., Hampshire, R., Clarke, L., Gibson, K., and Harris, P. (1972). An electron microscope technique for measuring the distribution of ribosomes and polysomes in cardiac ribosomal preparations. *J. Mol. Cell Cardiol.*, 4, 209.

Horne, R. W. (1965). Techniques for Electron Microscopy (Kay, D. H., Ed.), p. 311. Blackwell Scientific Publications, London.

Huston, R. L., Schrader, L. E., Honold, G. R., Beecher, G. R., Cooper, W. K., and Sauberlich, H. E. (1970). Factors influencing the isolation of a polyribosomal system from rat liver and its ability to incorporate (^{14}C)-amino acids. *Biochem. Biophys. Acta*, 209, 220.

Korner, A. (1958). Effect of hypophysectomy on the ability of rat liver microsomes in a cell-free system to incorporate radioactive amino-acids into their proteins. *Nature*, 181, 422.

Moroz, L. A. (1967). Protein synthetic activity of heart microsomes and ribosomes during left ventricular hypertrophy in rabbits. *Circ. Res.*, 21, 449.

Noll, H. (1969). Techniques in Protein Biosynthesis, Vol. 2. (Campbell, P. N., and Sargent, J. R., Eds.), p. 101. Academic Press, New York.

Norman, M. (1970). Separation with Zonal Rotors. (Reid, E., Ed.), p. 61. Wolfson Bioanalytical Centre of Univ. of Surrey, England.

Schreiber, S. S., Oratz, M., Evans, C., Silver, E., and Rothschild, M. A. (1968). Effects of acute overload on cardiac muscle mRNA. *Amer. J. Physiol.*, 215, 1250.

Takanami, M. (1960). A stable ribonucleoprotein for amino acid incorporation. *Biochem. Biophys. Acta*, 39, 318.

6. PARTICLE COUNTING OF VIRUSES

Mahlon F. Miller II

Department of Virology, The University of Texas, M. D. Anderson
Hospital and Tumor Institute, Houston, Texas

INTRODUCTION

Almost all virological observations take into consideration an estimate
of the number of virus particles involved. Techniques by which the concentration of virus particles in a preparation may be indirectly measured
include various bioassay methods (e.g., pock or plaque titrations) and
serological methods (e.g., hemagglutination or complement fixation
tests). However, the direct counting of virus particles, whether they are
complete virions or other entities recognizable as viral, is almost solely
within the province of electron microscopy. For over two decades, electron microscopists have been designing experiments which enable them to
calculate virus particles within certain limits of accuracy in a preparation.

There are several publications which review the complete literature on
virus particle counting studies (Isaacs, 1957; Sharp, 1963, 1965). Recently,
however, a number of new particle counting methods have been reported
which expand the general applicability of the virus particle counting procedures. This chapter is not intended to be a comprehensive review of all
the various counting methods ever tried, nor does it cover the various
applications of these methods. The purpose of this chapter is to provide
in some detail an explanation of the most commonly used methods for
enumerating virus particles in the electron microscope. It is hoped that
the electron microscopist, after reading the descriptions, will be able to
select the counting method which satisfies his research interests in terms
of practicality and accuracy desired.

GENERAL PRINCIPLES

The basic principles underlying all particle counting methods were established in 1949 with the publication of the spray droplet (Williams and Backus, 1949) and sedimentation (Sharp, 1949) counting procedures. Since that time, a number of methods have been devised which are essentially modifications of these procedures with noted improvements. The task is basically one of depositing on a specimen grid, from a known volume of virus suspension, an array of virus particles which are clearly and individually identifiable, are present in sufficient numbers to make enumeration feasible, and are truly representative of the population of particles in the original suspension. The various counting procedures may be classified according to the manner in which virus particles are deposited onto specimen grids and are here divided as follows: (i) methods in which polystyrene latex reference spheres are incorporated into virus preparations, (ii) methods which involve the direct sedimentation and counting of viruses, and (iii) methods which call for ultrathin sectioning of virus preparations.

In the reference particle methods, known concentrations of readily recognizable particles are mixed with the virus suspension and either sprayed or spread on the surface of filmed specimen grids. The average number of both virus and reference particles in randomly selected microscope fields is determined. From their ratio and the known concentration of reference particles, the concentration of virus in the original suspension is calculated. Bacteria, whose concentration can be determined with light microscopy in a counting chamber, have been used as reference particles by some investigators. However, most microscopists employ polystyrene latex (PSL) spheres as reference particles, since they are available in a variety of known sizes and concentrations.

Sedimentation counting procedures involve the direct sedimentation of virus particles from known volumes onto receiving surfaces such as glass or congealed agar. Sediments are then pseudoreplicated, metal shadowed or stained, stripped from the receiving surface, and picked up on specimen grids. From the average number of virus particles in microscope fields of known dimension, and with knowledge of the geometry of centrifuge cells from which they were sedimented, the concentration of virus in the original suspension is calculated. Some investigators have attempted to sediment particles directly onto filmed specimen grids. However, the preferred method is to first centrifuge virus particles onto an intermediate flat supporting substrate.

The thin section counting methods are essentially modifications of the sedimentation counting procedures. Virus particles are sedimented from a suspension either into the tip of a bottleneck embedding capsule or onto a flat-surfaced supporting membrane. After embedding, virus particles are counted in cross sections of the viral disks thus formed. Average counts in fields of known dimension are multiplied by either a volume or surface area factor to yield the total number of virus particles in the original suspension. One of the principal advantages of these methods is that viruses which are difficult to distinguish following metal shadowing or negative staining may be easily recognized in thin sections, so that crude or semipurified samples may be quantitated.

There is probably no single counting procedure which will fully satisfy all the requirements of an electron microscopist. While one method may desirably yield fast results, it may not provide the precision, accuracy, or sensitivity needed. The section on choosing a method will, hopefully, aid the microscopist in selecting a particle counting technique suitable for his studies.

LATEX REFERENCE PARTICLE METHODS

Polystyrene latex (PSL) spheres manufactured by the Dow Chemical Co., Midland, Michigan, are used in a variety of particle counting methods. Due to their dimensional uniformity (Backus and Williams, 1949), PSL spheres can readily be identified in the electron microscope when mixed with virus particles of a different size or shape. By mixing a known concentration of PSL spheres (CL) with a suspension containing virus particles to be counted, and applying this material to a filmed specimen grid, it is possible to determine the concentration of virus (CV) in the suspension:

$$\mathrm{CV} = \frac{\text{Av. no. virions per unit vol.}}{\text{Av. no. PSL spheres in same vol.}} \times \mathrm{CL}$$

The concentration of PSL spheres used in this mixture depends on the concentration of virus particles being counted and should be chosen so that the values are of the same order of magnitude.

Polystyrene Latex Sphere Concentration

To determine the concentration C of PSL spheres in a suspension it is necessary to know the mean particle diameter $2r$, the weight w, and density ρ of the dried latex:

$$C = \frac{3w}{4\pi r^3 \rho}$$

The density ρ of PSL spheres has been shown to be 1.05 g/ml by density gradient centrifugation. Most investigators report that PSL particle diameters as reported by the manufacturer are accurate to within the range of experimental error of the counting methods. The mean diameter of spheres may be checked by comparison with a reference standard such as a diffraction grating replica. While determining particle diameters, PSL sphere preparations may be checked for particle aggregation. The use of badly clumped PSL spheres for particle counting leads to inaccurate calculation of virus concentration (Watson, 1962a).

Spray Droplet Method

The method of spraying particulate suspensions onto specimen supports for examination in the electron microscope was introduced by Riedel and Ruska (1941). Utilization of the spraying method in combination with PSL reference spheres for virus particle enumeration was originated by Williams and Backus (1949). In this method the virus suspension mixed with a known concentration of PSL spheres is sprayed from a spray gun so that minute droplets of the mixture fall onto a filmed specimen grid. As suitably sprayed microdrops dry, their contents are deposited over discrete circular areas of a grid so that electron micrographs of individual droplets (2–20 μm diameter) can be taken. Since each droplet is a random sample of the total virus–PSL sphere suspension, virus concentration may be calculated from average counts of the number of virus particles and PSL spheres.

The use of nonvolatile buffer salts in the suspending medium should be avoided, since dried salts on the specimen screen may interfere with particle counting. Attempts to wash away salt deposits may result in the loss of portions of specimens and lead to inaccurate counts. Freshly prepared 0.15M ammonium acetate buffer adjusted to pH 7.5 with 0.3M ammonium carbonate may be used as the suspending medium for viruses which are not stable in aqueous suspension. It is recommended that serum albumin be added to the suspension to facilitate spreading of droplets and to outline their boundaries.

Initially, investigators utilizing this counting procedure employed metal shadowing to enhance contrast of specimens. Watson (1962a) combined the advantages of negative staining (Brenner and Horne, 1959) with the spray droplet counting procedure. Accordingly, 0.1 ml of virus suspension, 0.1 ml of PSL sphere suspension, 0.1 ml 2% PTA adjusted to pH 7.4 with 1N KOH, and 0.05 ml 1% bovine serum albumin are mixed and sprayed onto carbon-coated grids. Virus particles and PSL reference

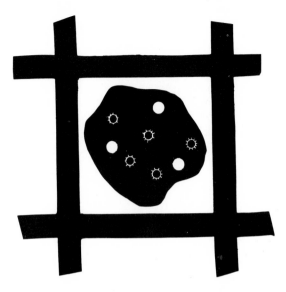

Fig. 6.1 A schematic diagram of a grid square with a dried-down droplet of PTA containing virus particles (structures with spokes) and PSL reference spheres (clear circles).

spheres in a dried-down droplet of PTA are shown schematically in Fig. 6.1. This combination is now probably the most widely used adaptation of the spray droplet technique. It requires less time to prepare specimens without the additional shadowcasting step, allows better resolution and more accurate identification of virus particles, and permits recognition and counting of virus particles in preparations which have not been so highly purified.

The spray droplet method can only be used for virus suspensions containing relatively large numbers of virions. Backus and Williams (1950) reported the limit of detection to be 10^9–10^{10} virus particles per milliliter.

Many investigators have designed their own spraying equipment. Various spraying devices were described by Haschemeyer and Myers (1972). One of the most commonly used devices is the commercial atomizer manufactured by the Vaponefrin Co., New York, N.Y. In spraying viruses it should be remembered that the formation of aerosols is an effective means of spreading infection. When pathogenic materials are sprayed, appropriate precautions should be taken. Horne and Nagington (1959) described an easily constructed spraying box which attaches to the Vaponefrin atomizer. Grids are enclosed during the spraying process and exhaust air passes through a washing chamber and to the atmosphere through the flame of a Bunsen burner (Fig. 6.2).

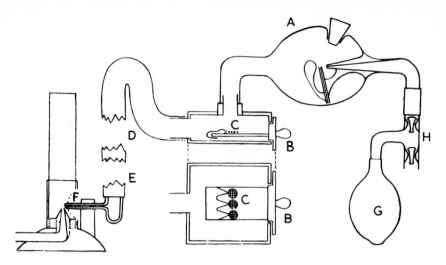

Fig. 6.2 Apparatus designed by Horne and Nagington (1959) for spraying infectious viruses. The Vaponefrin atomizer (A) was modified by adding a glass cone joint in place of its straight outlet tube. The atomizer was then fitted to a plastic chamber having a removable slide (B) to which specimen grids (C) were attached. The outlet tube from the chamber was attached to a wash bottle (D) and a reservoir (E). Following the reservoir, a small metal tube (F) permitted vapors to exhaust through the flame of a Bunsen burner. Air pressure to vaporize the virus suspension originated from the squeeze bulb (G) through the gas valve (H). (*J. Mol. Biol.,* **1,** 335, 1959.)

Agar Filtration Method

The agar filtration procedure divised by Kellenberger and Arber (1957) involves the spreading of virus suspension mixed with known concentrations of PSL spheres on collodion-coated agar blocks. Liquids and salts in the suspension subsequently diffuse through the collodion filters into the agar, leaving uniformly dispersed particulate components deposited on the collodion membrane. Membranes with deposited virus are subsequently affixed to specimen grids for particle counting. Details of the procedure are presented below.

1. Prepare agar plates by dissolving 1.5–3 g agar in 30 ml H_2O and 70 ml of the same fluid used to suspend the virus particles to be counted. Sterilize and pour 20 ml of this mixture into 9 cm Petri dishes. When the agar has gelled, clean agar and glass surfaces with distilled petroleum ether. The absorptive capacity of agar plates is increased by rendering them semidry. This is accomplished by drying plates in an oven until ~30% of the water content has evaporated. Dessication time can be standardized by weighing plates before and after drying.

2. Pour a thin layer of 0.2 to 0.4% parlodion in distilled technical grade amyl acetate over the surface of semidry agar plates. Invert plates and permit parlodion to dry for 6 hr.

3. Spread a few drops of virus suspension containing a known concentration of PSL spheres over the surface of the parlodion film with a smooth glass rod. Cover Petri dish and permit liquid to filter through parlodion film. Filtration is generally complete within 10 to 20 min, but will vary according to purity of virus suspension.

4. Following filtration, fixation is accomplished by inverting plates over the vapors of 40% formalin for 7 min. Small blocks of membrane and agar are subsequently cut out and dipped, membrane side up, into a solution of 2% $La(NO_3)_3$. The formalin and $La(NO_3)_3$ treatments facilitate separation of membranes from agar. Floating membranes are picked up from beneath on specimen grids and dried on filter paper.

5. Membranes may be metal-shadowed to enhance contrast. Five to ten electron micrographs are taken at a magnification consistent with virus particle recognition. Virus particles and PSL spheres are enumerated and the virus particle concentration in the original suspension is calculated from the ratio of spheres to particles, as in the spray droplet procedure.

The advantages of this method are derived from its simplicity, freedom from salt accumulation on membranes, and maintenance of virus particles in a medium of constant composition until they are fixed. However, particular care must be exercised in the preparation of specimens, especially in the casting of parlodion membranes. Membranes must be thin and free of holes. Kellenberger and Arber (1957) describe a lack of filtration by parlodion membranes which have been prepared from solutions of highly purified amyl acetate. The utilization of virus suspensions which contain high concentrations of serum or albumin should be avoided, since membranes tend to plug and thus filtration stops. The limit of detection of this method is comparable to the spray droplet procedure and works best at a virus concentration of $\sim 10^{10}$ per ml.

Lowered Drop Method

In the lowered drop method (Pinteric and Taylor, 1962) large drops of virus suspension slowly dry while resting on top of buffer-saturated sintered glass. Too rapid drying and the subsequent concentration effect at the periphery of drops is prevented. The method is suitable for virus suspensions containing proteins and salts. The presence of proteins in suspensions aids in spreading virus while salts are dialyzed away prior to drying.

1. Silver or gold grids (copper grids may react with acetate buffers) are placed on pieces of coarse sintered glass resting on the floor of a Büchner funnel. The funnel is filled via the bottom outlet with ammonium acetate buffer adjusted to pH 7.5 with ammonium carbonate.

2. A thin film, prepared from 0.2% Formvar in distilled ethylene dichloride, dried and stripped in the usual manner (Hayat, 1970) from a microscope slide, is floated on the surface of the buffer. Several 2–3 mm diameter drops of virus suspension mixed with a known concentration of PSL spheres are placed on the surface of the film and left for at least 30 min to allow for dialysis of salts through the film.

3. Buffer is drained out of the Büchner funnel and as the surface level falls, the film is guided with forceps so that drops come to rest on grids.

4. Grids, still resting on the pieces of sintered glass, are placed in a Petri dish. A watch glass containing NaOH pellets is placed beside the sintered glass and the top of the Petri dish is sealed with a rubber band. Sodium hydroxide absorbs CO_2 and prevents alteration of pH as drops dry. Drops on grids and buffer left in sintered glass evaporate in ~2 hr.

5. Dried grids may be metal shadowed to enhance contrast or may be negatively stained by soaking sintered glass in PTA before drops are completely dry. Virus concentration is determined by counting PSL spheres and virus particles in randomly selected fields. Fields are selected from one edge and across the diameter of the drop. Calculations are carried out as in the agar filtration or spray droplet methods.

The main advantages of this method are as follows: it may be used to enumerate virus particles in lower concentrations than in the two preceding PSL methods, and it can be used for virus samples containing relatively large amounts of protein. Pinteric and Taylor (1962) were able to estimate poliovirus concentrations as low as 5×10^7 per milliliter. It is recommended that serum albumin (0.01%) be added to highly purified virus suspensions to ensure uniform distribution of virus particles in dried drops.

Loop Drop Method

The simplest methods for estimating the number of virus particles in suspension involve the direct application of virus–PSL sphere mixtures to filmed specimen grids. Watson (1962b) reported a method for enumerating virus particles in relatively low concentration which involves the application of virus to filmed grids with a platinum wire loop. Virus suspension (0.1 ml) and PSL sphere suspension (0.1 ml), each estimated to contain ∼10^9 particles, are mixed with 0.1 ml of 0.5% PTA and 0.1 ml of 0.5% serum albumin. Relatively large drops of this mixture are applied to Formvar-coated grids with a loop and allowed to dry. Herpesvirus particles prepared by the loop drop method are shown in Fig. 6.3. Virus and PSL sphere counts are made directly from the microscope screen. Calculations are performed as in the preceding methods.

Particle counts by the loop drop method do not differ significantly from counts derived by the spray droplet or agar filtration methods (Watson

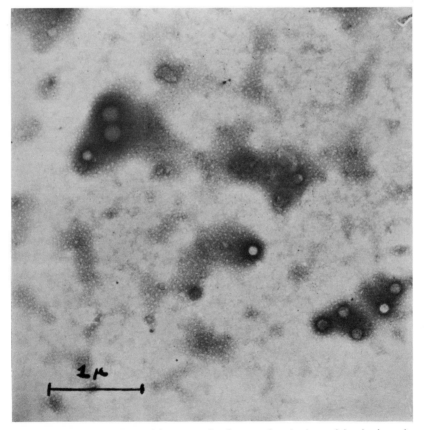

Fig. 6.3 Herpesvirus particles and latex particles (larger spheres) prepared by the loop drop counting procedure. (Courtesy of D. H. Watson.)

et al., 1963). The presence of proteins in the suspending medium permits uniform drying of drops. Salts in the suspending medium are apparently sufficiently dilute not to be troublesome. Moreover, PTA probably stabilizes the morphology of virus and renders it more easily recognizable.

An additional single-drop method which combines advantages offered by two other techniques was introduced by Geister and Peters (1963). These authors applied a drop of virus–PSL sphere mixture to a filmed grid floating on the surface of water. The filmed grid was pretreated with 0.005% serum albumin to reduce aggregation of particles caused by hydrophobic properties of the film, and the drop of virus–PSL sphere mixture was placed in the center so that it did not reach the edges. Evaporation was prevented by covering the dish. As the drop dialized through the film, particulate components were deposited on the surface. When virus particles and PSL spheres were counted in fields corresponding to complete grid square openings, constant PSL sphere to virus particle ratios were achieved, and successful counting in samples containing as few as 10^5 virus particles per ml was reported.

SEDIMENTATION METHODS

Sedimentation counting methods are, in general, more complicated and time consuming than the reference particle methods. However, they offer the distinct advantage of greater sensitivity and samples containing relatively few virus particles may be counted. In contrast to the spray or drop methods, PSL reference spheres need not be used to facilitate virus titer calculations. Instead, virus particles from known volumes of virus suspensions are sedimented directly onto a receiving surface. Pseudoreplicated, metal-shadowed, or stained virus particles are subsequently counted in the electron microscope, and counts are related to the starting volume.

In the first sedimentation method (Sharp, 1949), collodion-coated coverslip glass served as the receiving surface for sedimented virions. The method was later modified (Sharp and Beard, 1952) to take advantage of the absorptive properties of agar as a receiving surface. In this method, much of the excess protein from virus suspensions may wash away and remaining buffer salts diffuse into the agar, leaving particulate components adhering to the surface. A special rotor has been designed (Sharp and Overman, 1958) (see Fig. 6.4) for sedimenting virions onto an agar surface, and the rotor may be purchased from Ivan Sorvall Inc., Norwalk, Conn. The cells of this rotor are designed so that convection in the suspending medium is minimized. The walls of cells are trapezoidal and cor-

Fig. 6.4 The Servall-Sharp SU rotor designed for sedimenting virus particles for counting. (Courtesy of Ivan Sorvall, Inc.)

respond to the radii of the rotor, so that virus particles are evenly sedimented over the agar surface (Fig. 6.5).

To make a particle count by the agar sedimentation procedure it is necessary to choose a microscope magnification at which the virions may be easily recognized and to estimate a range of virus dilutions at which microscope fields delineated by this magnification will contain a suitable number of virions for counting. Since there are eight cells in the rotor under discussion, a wide variety of dilutions may be sedimented at the same time and particle counts may be made on only those preparations containing adequate numbers of virions. Details for preparing specimens are as follows.

1. Partially assemble centrifuge cells by placing one lucite disk, one gasket, and a sector-shaped piece into metal shell.

2. Insert an agar block, $\sim$2 mm thick by 1 cm^2, so that it lies flat at the base of the sector. Agar pieces are prepared by heating 2 g of nutrient agar with 98 g of buffer and pouring 20 ml of solution into 10 cm Petri dishes. After hardening, pieces of appropriate size are cut out with a spatula and placed in sectors.

3. Virus suspensions are pipetted to fill cells until the surface is slightly convex. The second gasket and lucite disk are carefully placed on top of the cell so that all air bubbles escape as the lock ring is screwed into place.

4. Assembled cells are placed in the rotor with agar surfaces outward

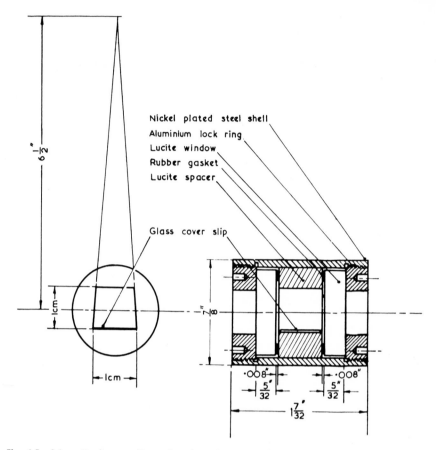

Fig. 6.5 Schematic diagram illustrating the construction of centrifuge cells devised by Sharp (1949) for the convection-free sedimentation of virus particles. (By permission of *Soc. Exptl. Biol. Med.* **70,** 55, 1949.)

and the rotor cover is tightened in place. A duration and speed of centrifugation are chosen which correspond to complete sedimentation of virus. Accordingly, influenza virus will sediment from physiological saline in ~30 min at 15,000 rpm, and vaccinia virus will sediment in ~15 min at 10,000 rpm.

5. After sedimentation, the tops of cells are removed, supernatant fluid is pipetted off, and agar pieces with adhering virus are removed with a spatula. Agar pieces are placed on microscope slides and set on edge to drain for ~3 min.

6. Pseudoreplicas of virus particles are made by flooding the agar surface with 1% parlodion in amyl acetate and permitting it to drain dry for ~5 min. The parlodion film thus formed, which contains the virus particles (pseudoreplica), is floated off onto the surface of clean water.

7. Replicas are inverted and mounted on specimen grids for examination in the electron microscope. Although it is not essential, metal shadowing of replicas enhances contrast and makes virus particles much easier to count. Virions may also be stained (see below).

8. Virus particles are counted in a number of microscope fields and averaged. The number of fields which should be counted is dependent upon the desired accuracy and the observed uniformity of virus distribution, i.e., amount of virus aggregation apparent in micrographs. The number N of particles per milliliter is then calculated from the equation

$$N = \frac{nM^2D}{0.91Ah}$$

where n is the average number of virus particles in a known area A (field size in cm²), M is the magnification, h is the height of the liquid in the centrifuge cell (subtract agar thickness), and D is the dilution factor. The factor 0.91 appears in the denominator to correct for the taper of centrifuge cell walls.

The above method has been used to count a wide variety of types of virus particles in the electron microscope with greater sensitivity than generally can be obtained by reference particle methods. Vaccinia virus particles prepared for counting by the agar sedimentation procedure are shown in Fig. 6.6. The accuracy of counts derived by the agar sedimentation method is comparable to the spray droplet method, and samples containing as few as 10^6 virions per milliliter may be counted (Sharp, 1965).

The original sedimentation counting method of Sharp has been modified by many investigators to suit varying experimental conditions. Attempts have been made to simplify sample preparation by sedimenting virions from suspensions directly onto filmed specimen grids (Crane, 1944; Overman and Tamm, 1956; Sharp, 1960; Smith and Melnick, 1962; Ball and Harris, 1968, 1972). Although satisfactory counts were derived in many of these studies, it was reported that the films often deform or tear during centrifugation. In addition, virus particle distribution is not uniform, as particles tend to concentrate near the periphery of depressed centers of grid squares (Smith and Melnick, 1962). Special centrifuge tubes suitable for sedimenting virus particles onto filmed grids are available from E. F. Fullam, Inc., Schenectady, New York. Ball and Harris (1972)

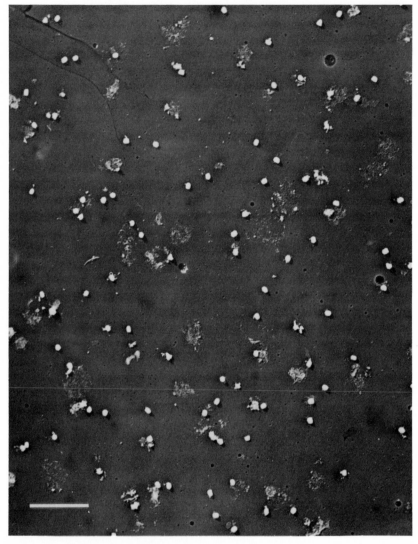

Fig. 6.6 Metal shadowed vaccinia virus particles prepared by the agar sedimentation procedure. Particles are uniformly distributed and display no aggregation. The bar equals 2 μm. (Courtesy of D. G. Sharp.)

utilized disposable nylon tubes fabricated by United-Carr, Knoxville, Tennessee, to sediment polio virus, Qβ bacteriophage, and latex spheres onto 400 mesh Formvar carbon-coated copper grids. Due to the very large variations in particles encountered per grid area, they concluded that an inordinate number of grid areas must be counted to obtain a mean count at the 95% confidence level.

When crude extracts of virus-infected tissues or tissue culture fluids are sedimented onto a receiving surface for virus particle counting, it is often difficult to recognize virions among cellular debris and make accurate counts. One method for reducing this recognition problem is to dilute the extract to the point where a minimum amount of nonviral debris is present, yet a sufficient number of virions are present to yield significant particle counts. Crude extracts of large virions such as vaccinia have been reliably counted in this manner (Overman and Tamm, 1956; Sharp and Overman, 1958). Another approach is to digest virus extracts with trypsin and chymotripsin in order to reduce most cellular debris to unsedimentable dimensions and at the same time aid in the dispersion of particles (Smith and Sharp, 1960; Smith and Benyesh-Melnick, 1961; Smith and Melnick, 1962).

A modification of the agar sedimentation procedure which improves the ability to recognize virions is the substitution of staining for metal shadowing in the pseudoreplication process (Rhim et al., 1961; Smith and Melnick, 1962). In this modification, virus particles are sedimented onto the agar surface and coated with collodion as before. However, the pseudoreplica is removed by flotation onto the surface of either 0.25–0.5% uranyl acetate or potassium phosphotungstate. After 10–20 sec the stained pseudoreplica is picked up on a grid and dried. Pseudoreplicas of virions which have been positively (uranyl acetate) or negatively (phosphotungstate) stained exhibit greater contrast and better structural preservation than pseudoreplicas which have been metal shadowed. Smith and Melnick (1962) used specific nucleases and the above stains to differentiate between RNA and DNA viruses. Problems associated with virus recognition may also be alleviated by counting virus preparations which have been thin-sectioned (see the section on Thin Section Methods in this chapter).

Since the agar sedimentation method was developed, several investigators have endeavored to devise additional centrifuge cells which are suitable for preparing specimens for particle counting. The trend has been to make cells for the swinging bucket type of centrifuge rotor. No commercially fabricated centrifuge cells for the swinging bucket rotor are presently available which were specifically designed for preparing specimens for agar sedimentation counting.

Smith and Benyesh-Melnick (1961) described flat-bottomed lucite cells for the Spinco SW25.1 rotor. These cells consist of a cylindrical chamber 19 mm deep and 19 mm in diameter into which a hollow cylindrical insert is placed. Agar disks ~13 mm in diameter cut with a cork hole borer are placed at the bottom of the cell. One milliliter of appropriately diluted virus suspension is pipetted into the cell and virions are sedimented onto the agar surface. Particle counts on polyoma virus prepared with these cells revealed a radial spread of virions on the flat agar surface due to the cylindrical rather than sectoral shape of centrifuge cells. The authors reported that the uneven spread of virions resulted in a dilution of ~7%, and a correction factor of 1.07 was used. Another correction factor of 1.07 was used to correct particle counts, since it was found that only 93% of the total virus was recovered during the pseudoreplication process. Polyoma virus particle counts with a precision of ~16% were obtained by this method.

From the previous discussion it is clear that success with sedimentation counting methods is dependent upon the construction of the centrifuge cells employed. Centrifugation of suspended material in a conical shaped cell will result in the deposition of more material at the periphery of the receiving surface than at the center. Although less pronounced, these so-called "edge effects" will also occur when suspended material is centrifuged in a cylindrical cell. The centrifuge cells designed by Sharp for use in Sorvall rotors are wedge shaped such that the base of the cell is larger than the top of the cell. The walls of these cells are so designed that they correspond to the radii of the rotor. Edge effects are, therefore, not encountered.

Mathews and Buthala (1970) have designed a centrifuge cell for the swinging-bucket-type rotor, which may be seen in various stages of assembly in Fig. 6.7. A single cell is shown schematically in Fig. 6.8. Each stainless-steel cell is composed of three parts: (A) a threaded handle for removing assembled cells from the centrifuge tube following centrifugation, (B) a cylindrical body with a lumen which conforms to the shape of a conic frustum, and (C) hemispherical base with a flat-topped boss. The base and body of the cell fit into Spinco stainless-steel centrifuge tubes for the SW39 rotor (Beckman Instruments, Inc.). Because of the extra weight of cells, it is suggested that centrifugation speeds be at a maximum of 20,000 rpm to avoid exceeding the safety factor of rotors.

The above centrifuge cells have been used by Mathews and Buthala (1970) for counting virus particles of a variety of types as well as known concentrations of polystyrene latex spheres. Due to previous reports that some virions may be lost during the pseudoreplication process (Sharp, 1960; Smith and Benyesh-Melnick, 1961; Sharp, 1965), a receiving surface of carbon-coated glass disks was chosen. Since the lumen of the cell is shaped like an inverted cone and is not sector-shaped, particular care must be taken to count only virus particles which are deposited at the center of the glass disk. The procedure is as follows:

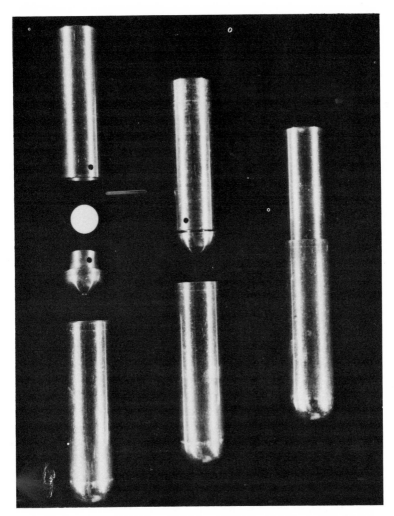

Fig. 6.7 Centrifuge cells designed by Mathews and Buthala (1970) for uniformly sedimenting virus particles onto a flat surface in the swinging-bucket-type rotor. (Courtesy of J. Mathews.)

1. No. 3 thickness 8 mm diameter cover glasses (specially ordered from Corning Glass Works, Corning, N.Y.) are cleaned by soaking for several hours in acetone followed by absolute ethanol, and dried in a dust-free chamber.

2. Disks are coated with carbon and placed in centrifuge cells. Virions are sedimented onto the carbon surface and after centrifugation the specimen is fixed for 15 min in OsO_4 vapors.

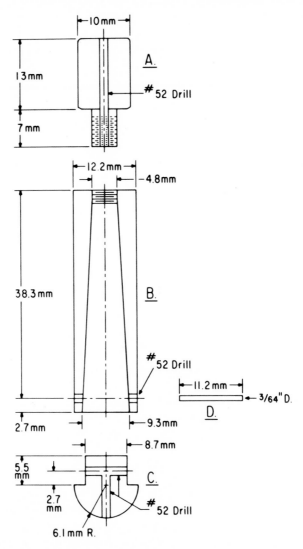

Fig. 6.8 Schematic diagram of one of the centrifuge cells shown in Fig. 6.7. (Courtesy of J. Mathews and by permission of *J. Virol., 5, 599, 1970.*)

3. Specimens are washed briefly in distilled water and dehydrated in a graded series of alcohols. Disks are subsequently air-dried and shadowed with platinum.

4. After shadowing, disks are scratched with a sharp tool so that carbon films with deposited virus are divided into quadrants. Disks are then im-

mersed for 1–5 min in 25% potassium hydroxide, removed, drained, and again lowered into the potassium hydroxide solution at an acute angle. In this way the film is stripped from the glass surface and left floating on the liquid surface.

5. Films with deposited virus are picked up from beneath on a fine-mesh screen and transferred successively to the surfaces of water, 0.2N hydrochloric acid, and two additional water baths.

6. Portions of washed films corresponding to the centralmost area of glass disks are picked up from beneath on 200 mesh copper grids. Grids are subsequently drained, air-dried, and examined in the electron microscope.

Particle counts may be made either directly from the microscope screen or from pictures of various fields. Fig. 6.9 shows a typical field of Coxsackie-virus A-21 prepared by this procedure. Virus titers are determined from average counts per field as in the other sedimentation methods. Cal-

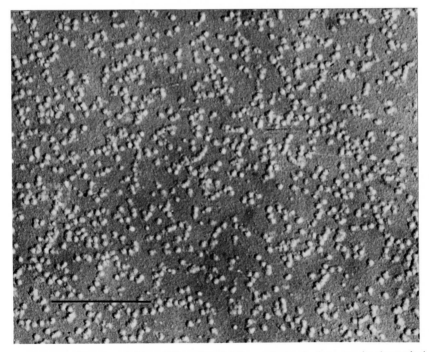

Fig. 6.9   Coxsackie-virus particles, A-21, sedimented in a swinging bucket rotor by the method of Mathews and Buthala (1970). The bar equals 1 μm. (Courtesy of J. Mathews and by permission of J. Virol., 5, 600, 1970.)

culations may be made from the equation previously cited (Sharp, 1965). However, a taper correction factor of 0.869 rather than 0.91 is used because of the difference in cell geometry.

All of the counting methods described up to this point require reasonably pure virus preparations so that virions may be easily distinguished from nonviral material. Strohmaier (1967) designed special centrifuge cells in which virions may be purified by rate-zonal centrifugation in a density gradient, and at the same time be sedimented directly onto coated metal grids for particle counting. Two types of centrifuge cells designed for this purpose are shown in Fig. 6.10. The first type, which may be purchased from E. F. Fullam, Inc., Schenectady, New York, is shown disassembled in Fig. 6.10A, and assembled in Fig. 6.10B. The second type of cell is shown assembled in Fig. 6.10C. Both types of cells may be used in rotors such as the Spinco SW25 with 1×3 in. swinging buckets.

The first type of centrifuge cell is composed of a base a with a central shaft which supports segments b, c, d, e, f, and g. The segments c, d, e, and f, have two holes drilled opposite one another and an O-ring is positioned in a recess at the bottom of each hole. Segments c, d, and e also have two depressions in the top surface which are located at $90°$ angles to holes. Segment b has depressions in all four positions. The top segment, g, is used to close off holes after loading the cell, and a knurled nut, h, holds all segments together. Depending on the alignment of the various segments, a channel through all sections with the bottom closed off by the lowest segment may be formed, or the hole in each segment may be closed off by the segment immediately beneath. A Formvar–carbon-coated-nickel grid supported by a bead of mercury may be placed either in the depression of each segment or only in depressions of the lowest segment. Alternatively, coated grids of any type may be supported by silicon rubber disks which float on top of a drop of heavy oil (density 1.9 g/ml, FC 43, 3M—Minnesota Mining and Manufacturing Co.) in each depression. Nickel grids must be used with mercury drops because nickel does not amalgamate with mercury.

The second type of centrifuge cell (Fig. 6.10C) is composed of the same segments a, b, g, and h. However, the middle segments are replaced by a single cylinder having only one hole, which may be turned over each depression in segment b. The hole in this cylinder has a sectoral shape which yields uniform sedimentation of particulate components onto grids.

After cells are assembled, the various segments are filled with solutions of decreasing concentrations to form the desired gradients. Gradients may be formed from any of the common density gradient reagents, such as sucrose or glycerol. However, Strohmaier (1967) recommends use of solutions which leave no residue on evaporation. Accordingly, mixtures of

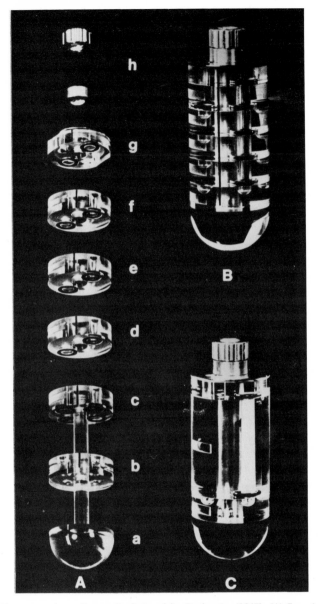

Fig. 6.10 Two types of centrifuge cells designed by Strohmaier (1967). (A) Type I disassembled and consisting of parts *a–h*. (B) Type I assembled. (C) Type II cell assembled. See text for details. (Courtesy of K. Strohmaier and by permission of *J. Virol.* **5,** 1075, 1967.)

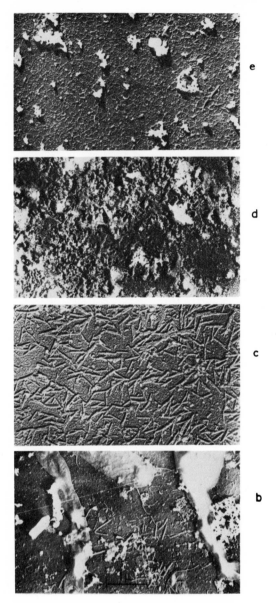

Fig. 6.11 Electron micrographs of platinum-shadowed density gradient fractions of a plant extract containing TMV particles. The letters b–e correspond to segments of the Strohmaier (1967) cell shown in Fig. 6.10A. Most of the rod-like TMV particles were found at the level of segment c. (Courtesy of K. Strohmaier and by permission of the *J. Virol.*, **5**, 1078, 1967.)

heavy and normal water ($D_2O–H_2O$) or glycerol and water are most satisfactory. To maintain certain ionic or osmotic conditions consistent with maintenance of virus morphological integrity, volatile salts such as ammonium acetate may be added.

Virus particle suspensions are layered over the top of gradients and the top segment, g, is turned to close off the holes. Two centrifugations are performed. During the first run, the components of the test suspension are separated according to their sedimentation velocities. Then the centrifuge is stopped and sections of the centrifuge cell are turned so that coated specimen grids are inserted into the path of sedimenting components in the gradient. During the second centrifugation run, virus particles or cellular components trapped between segments are sedimented onto grids. Following centrifugation, the supernatant fluid is withdrawn and grids are removed for electron microscopic examination. Virus particle titers are calculated with the same equation used in the other sedimentation methods. However, a taper correction factor does not appear in the denominator when cylindrical cells are utilized.

Figure 6.11 shows an example of results derived by the Strohmaier method. A crude plant extract containing tobacco mosaic virus (TMV) was sedimented in the special centrifuge cells on a $D_2O–H_2O$ gradient. Components of the extract were sedimented to the various levels of the gradient according to their sedimentation velocities. Segments b–e of Fig. 6.11 correspond respectively to segments b–e of the centrifuge cell depicted in Fig. 6.10A. Primarily, bacteria and cellular debris were sedimented to the level of segment b; the rodlike TMV particles may be seen at the level of segment c; dense granules are found in segment d; and salt crystals with a fibrous component are found in segment e.

THIN-SECTION METHODS

The methods described for counting virus particles in the electron microscope involve the enumeration of metal-shadowed or negatively stained virus particles which have been suitably deposited on electron-transparent films. All of these methods require the use of reasonably pure virus particle suspensions in order to minimize difficulties encountered in recognition and counting of virions among nonviral material. Many investigators have found that reliable recognition of some virus particles may be accomplished only in thin-sectioned specimens in which characteristic viral internal structures may be seen. Recently, two methods were reported for counting particles in thin sections of sedimented virus preparations (Gehle and Smith, 1970; Miller et al., 1972; Miller et al., 1973). Both methods permit reliable recognition of virus particles among cellular debris and retain the sensitivity and precision obtainable by the other sedimentation counting methods.

In the particle counting method devised by Gehle and Smith (1970), aliquots of virus suspensions are prepared for thin-section electron microscopy by pelleting, fixing, and embedding in BEEM bottleneck capsules. Bottleneck capsules may be purchased from a number of suppliers, and only those which are uniform and free of flaws should be used. Adaptors to hold BEEM capsules in Spinco ultracentrifuge rotors may be fabricated from Lucite (Smith and Gehle, 1969) or liquid casting plastic (Gehle and Smith, 1970).

Virus particle pellets are cut from the tip of bottleneck blocks (Fig. 6.12A), rotated 90° and re-embedded with the same batch of resin in silicon rubber molds (Fig. 6.12B). Virus particle distribution and pellet height Hp are determined by examining thin sections of the flat-embedded pellet. Pellet diameter Dp may be ascertained either by measuring the in-

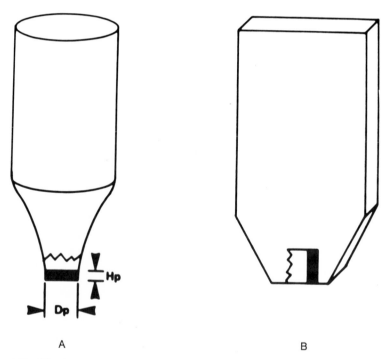

A   B

Fig. 6.12   (A) Diagram of a bottleneck capsule with a pellet of height Hp, and diameter Dp. (B) The pellet is cut out and flat-embedded. (Courtesy of Dr. K. O. Smith and by permission of Soc. Exptl. Biol. Med., 135, 490, 1970.)

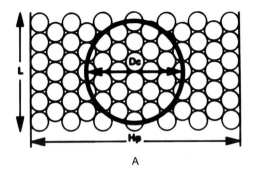

A

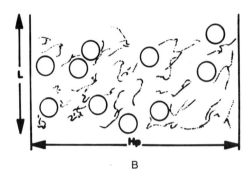

B

Fig. 6.13 The distribution of purified virus (A) and unpurified virus (B). Virus particles are counted either in a circular area with a diameter Dc or in a rectangular area L × Hp. (Courtesy of K. O. Smith and by permission of *Soc. Exptl. Biol. Med.*, **135**, 491, 1970.)

side diameter of the bottleneck capsule or by measuring the outside diameter of the pellet with a micrometer disk, and was found by Gehle and Smith (1970) to be 8.0×10^{-2} cm. Pellet volume Vp is calculated from the equation

$$Vp = \pi(Dp/2)^2 Hp$$

In pure samples, virus particles are counted in circular areas of electron micrographs, as shown schematically in Fig. 6.13A. The volume Vd of the disk delineated by this circle in which virus particles are actually counted may be determined from the equation

$$Vd = \pi(Dc/2)^2 S$$

where Dc is the diameter of the circle and S is the section thickness. Section thickness may be estimated to within 10 or 20 nm according to the interference color (Peachey, 1958). From the average number of particles counted per circular field (N) the total number of virus particles (T) in the entire pellet may then be calculated from the formula

$$T = NVp/Vd$$

Nonviral material is present in many samples, and virus particles may not be uniformly distributed throughout the height of the pellet. Since the average number of virus particles per field (N) must represent the average value for virus particles present throughout Hp, in impure pellets it is more accurate to count particles in an area $L \times Hp$ (Fig. 6.13B). The total number of virus particles in the entire pellet (T) may then be calculated from the formula

$$T = \frac{N\pi(Dp/2)^2}{LS}$$

From the total number of virus particles in the pellet the concentration of virus in the original suspension may then be calculated per unit volume as desired.

It is actually desirable to perform virus particle counts by the pelleting method on impure samples. Highly purified samples tend to lack coherence and may become disrupted during fixation, dehydration, and embedding. Should this be a problem, washed erythrocytes may be pelleted on top of viral pellets, as shown in Fig. 6.14, to hold the sample in place.

In the method devised by Miller et al., (1973), virus particles are sedimented onto membrane filter disks. Virus particles are then counted in ultrathin cross sections of the filter disk with adherent virus. Sediments so thin that they are not visible to the unaided eye may be counted, only a single embedment is required, and impure samples are easily quantitated. The apparatus used to prepare specimens for particle counting is shown in Fig. 6.15. Details of the procedure are given below.

Type VF (25 nm) Millipore filter disks of 4 mm diameter are punched from larger filters supported by a thin rubber mat using a piece of sharpened stainless steel tubing (4 mm i.d.). Disks are next placed glossy side up on top of reusable flattened supports made of epoxy embedding media in $\frac{3}{16}$ in. $\times$ $1\frac{5}{8}$ in. cellulose nitrate centrifuge tubes, as shown in Fig. 6.16. The use of a plunger made from the upper portion of a disposable 1 ml glass pipette aids in the insertion of epoxy supports and filter disks into the small centrifuge tubes. It is more convenient to sediment virus parti-

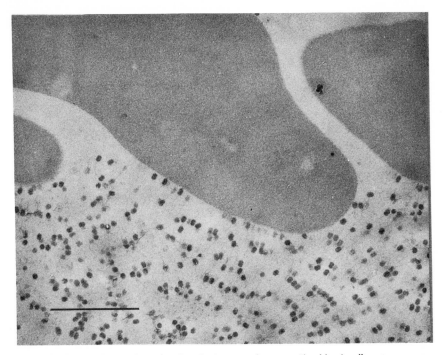

Fig. 6.14 T4 phage particles capped with sheep erythrocytes. The blood cells act as a protective plug to prevent virus pellets from being dislodged during processing. The bar equals 1 μm. (Courtesy of K. O. Smith and by permission of *Soc. Exptl. Biol. Med.*, **135**, 493, 1970).

cles onto the glossy surfaces of filter disks, since they have been found to be less irregular than the dull surfaces.

Virus particles from a known volume of suspension are next sedimented onto filter disks in a Spinco SW39L, SW50L, or SW50.1 ultracentrifuge rotor. The nitrocellulose tubes with epoxy supports in position hold ~0.6 ml of suspension. Small tubes and adaptors to hold them in centrifuge buckets may be purchased from Beckman Instruments, Palo Alto, California. Following centrifugation the supernatant fluid is carefully removed and replaced with glutaraldehyde fixative. After 1 hr at 4°C the centrifuge tube is slit down one side with a razor blade and spread open so that the filter disk with sedimented virus can be removed. Specimens are subsequently washed, post-fixed in osmium tetroxide, dehydrated, and flat-embedded in silicon rubber molds. Isopropanol and toluene are used in place of ethanol and propylene oxide for dehydration and clearing, because the latter two agents cause swelling and curling of Millipore filter disks.

Fig. 6.15 Apparatus used to sediment virus particles onto membrane filter disks. Six adaptors which hold the small tubes in the buckets of Beckman ultracentrifuge rotors are shown. Also shown is a cutter for punching out membrane disks and plastic supports made of Epon embedding media. The loading of one of the small tubes with plastic support and filter disk is demonstrated.

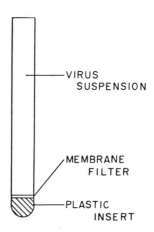

VIRUS
SUSPENSION

MEMBRANE
FILTER

PLASTIC
INSERT

Fig. 6.16 Schematic diagram of a $\frac{3}{16}$ in. $\times$ $1\frac{5}{8}$ in. cellulose nitrate centrifuge tube with plastic support and filter disk in position. The tube is filled with a suspension containing the virus to be counted. (By permission of *J. Gen. Virol.*, **21**, 59, 1973.)

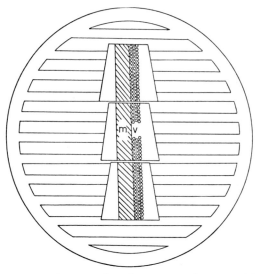

Fig. 6.17 Schematic diagram illustrating how thin sections of virus particles (v) which were sedimented onto membrane disks (m) are mounted on parallel wire grids (Miller *et al.*, 1973). (By permission of *J. Gen. Virol.*, **21**, *59*, 1973.)

Epoxy blocks in which filter disks have been embedded are oriented in the microtome so that sections can be cut perpendicular to the plane of the disk. As shown in Fig. 6.17, sections of filter disks are mounted perpendicular to grid bars of parallel wire grids. Specimens are subsequently stained, stabilized with evaporated carbon, and examined in the electron microscope. A portion of a field of murine sarcoma virus particles (MSV-SD, Soehner and Dmochowski, 1969) prepared in this way is shown resting on a filter disk for particle counting in Fig. 6.18.

To calculate the total number of virus particles sedimented onto the surface of the filter disk, the average number of particles per field between grid bars is determined and multiplied by a surface area factor. The length of one of these fields is limited by the lowest magnification consistent with virus particle recognition and can be measured against a diffraction grating replica. Virus particles are usually easy to recognize if they are photographed at a magnification of 1,500–2,000×, and negatives are either projected onto a screen or viewed with a dissecting microscope (Fig. 6.19). The lower the magnification at which specimens are photographed, the larger are the field size and the portion of the filter disk sampled. Particle counting is simplified if a grid of fine lines is superimposed on top of projected negatives.

The fraction of the total surface area of the filter disk which can be

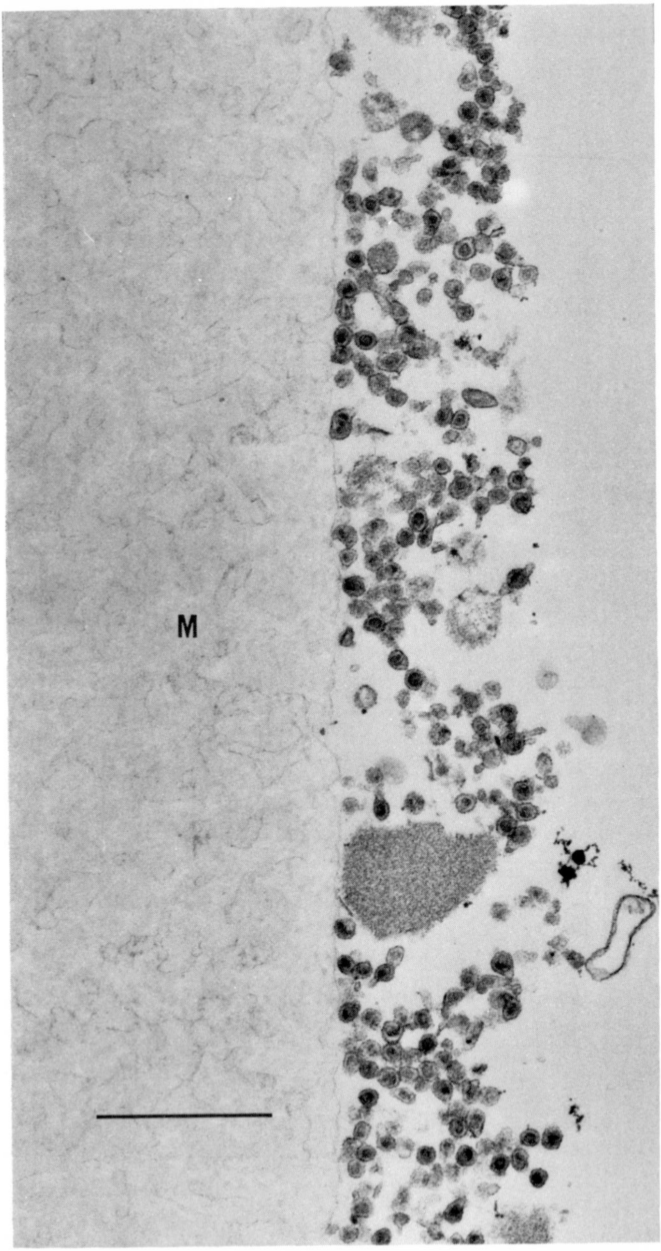

Fig. 6.18 Portion of a thin section of murine sarcoma virus particles (MSV-SD) resting on a filter disk (M) for particle counting. Virus particles are well preserved and easy to distinguish among cellular debris. The bar equals 1 μm.

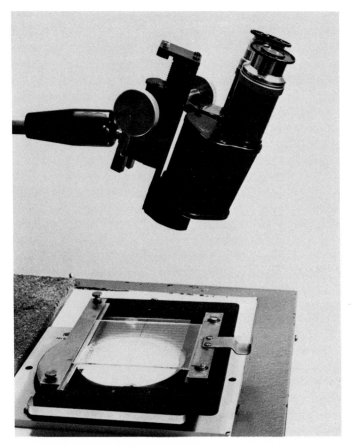

Fig. 6.19 A binocular dissecting microscope mounted above a movable illuminated stage for counting virus particles in negatives. A grid of fine lines placed over negatives aids in counting particles.

viewed in a single electron micrograph is limited by the length of the field at a given magnification and the thickness of a thin section. To determine this fraction, the investigator must estimate the portion of a virus particle which needs to lie within a given thin section to enable reliable recognition. Miller *et al.* (1973) estimated that at least one-half of an adenovirus particle, one-third of an oncornavirus particle, and one-quarter of a vaccinia virus particle should be present in a thin section to allow for virus recognition. Since thin sections of virus particles always include some particles which lie only partially in the section, a given thin section effectively samples a portion of the specimen slightly thicker than the section itself.

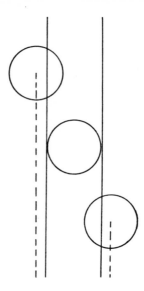

Fig. 6.20 A diagram showing how the effective section thickness may be estimated. Virus particles whose centers lie between the dotted lines (effective section thickness) would lie sufficiently within thin sections so that they could be recognized and counted. (By permission of *J. Gen. Virol.*, **21,** 60, 1973.)

Determination of the "effective section thickness" is schematically illustrated in Fig. 6.20 which shows virus particles having diameters approximately equal to the section thickness. If it is estimated that one-third or more of the particle diameter must extend into the section for reliable recognition to be made; it can be seen that particles whose centers lie outside the dotted lines would not be recognized or counted. All particles whose centers lie between the dotted lines would be counted. The effective section thickness T' is calculated from the formula

$$T' = T + D_v - 2fD_v$$

where T is the actual section thickness, D_v is the particle diameter, and f is the estimated fraction of the particle necessary for recognition. Although thin sections tend to vary somewhat in thickness (Peachey, 1958; Hayat, 1970), error in estimates of actual section thickness T can be minimized by averaging particle counts on fields from several different thin sections displaying the same interference color. Miller *et al.* (1973) used effective section thicknesses T' of 0.09 μm for adenovirus, 0.123 μm for oncornavirus, and 0.215 μm for vaccinia virus. Vaccinia particles were treated as if they were 0.25 μm diameter spherical particles in calculations.

The average area of fields in which virus particles are counted is, therefore, determined by multiplying the effective section thickness T' by the length L of the field examined. The total number of particles sedimented onto a filter disk, V_m, is then calculated from the total surface area of the disk and the average effective area of a field

$$V_m = \frac{\pi(D/2)^2 V_f}{LT'}$$

where D is the diameter of the filter disk and V_f is the average number of virus particles counted in fields of length L and thickness T'. From V_m the concentration of virus in the original suspension may be calculated per unit volume as desired.

CHOOSING A METHOD

Since virus particles may be enumerated by a variety of methods, the electron microscopist will need to select the method which most nearly satisfies his particular research interests. The choice of method will depend on a number of factors. Perhaps the most obvious consideration is the availability of particle counting apparatus. If a suitable centrifuge is not available, the microscopist will be limited to one of the reference particle methods. However, most modern research facilities are equipped with at least one high-speed centrifuge, and by either purchasing or fabricating centrifuge cells suitable for particle counting it is possible to count virus particles by any of the described methods. How much of an investment the microscopist will want to make in special particle counting apparatus will depend upon the counting sensitivity required and the frequency with which he expects to perform counts.

Probably the simplest, quickest, and least expensive method for estimating the number of virus particles in a suspension is the loop drop counting procedure (Watson, 1962b). In this method a single drop of a suspension containing virus particles to be counted and a known number of PSL spheres is applied to a filmed specimen grid and permitted to dry. Although satisfactory results have been obtained by this method, a word of caution is in order. As drops or droplets of a suspension dry, their contents, especially smaller particles, tend to concentrate near the periphery. Since virus particles and PSL spheres can be enumerated in only a fraction of a large drop, the ratio of virus to PSL spheres in a given microscope field may not be representative of the ratio in the original suspension. The addition of serum albumin to virus suspensions reduces these edge effects but may not completely eliminate them. Therefore, the investigator interested in precise measurements will need to scan such preparations carefully to be certain that virus particle distribution is uniform.

When the various particle counting procedures were devised, particular attention was devoted to obtaining preparations in which virus particles were uniformly distributed on microscope grids. The influence of the shape of centrifuge cells used for sedimentation counting on particle distribution has already been described. Ideally, sector-shaped cells should be used in which the walls conform to the radii of the centrifuge rotor (Sharp, 1949). However, the tendency for sedimented particles to be concentrated at the periphery of a cylindrical tube decreases as the ratio between tube diameter and rotor radius approaches zero. Therefore, error resulting from the use of a cylindrical tube of small diameter, such as those available from Beckman Instruments (Miller *et al.*, 1973), is probably negligible.

Another factor which may influence the uniformity of particle distribution is the state of aggregation of particles in the starting suspension. Watson (1962a) has shown that some preparations of PSL spheres are badly clumped and that their tendency toward aggregation must be considered if maximum precision is to be obtained. Some viruses, such as avian myeloblastosis virus (Sharp and Buckingham, 1956), influenza virus (Morgan *et al.*, 1956), and oncornavirus, which are gradually released from cells, show little tendency toward aggregation. However, other viruses, such as herpes and vaccinia, which must be released from infected cells either by freeze-thaw procedures or by sonic lysis, have been shown to be highly aggregated (Sharp, 1965 and 1968).

The aggregation of vaccinia virus particles following sonic lysis increases with time if suspensions are permitted to stand. Figure 6.21A shows well dispersed vaccinia virus particles diluted for count by the agar sedimentation procedure immediately following sonic lysis (20,000 cycles per second) from Earle's L cells. When such crude lysates are allowed to stand, even for as little as 1 hr following ultrasonic treatment, small aggregates of particles appear, as shown in Fig. 6.21B. Upon standing for longer periods of time, aggregates composed of several hundred vaccinia virus particles may be formed. While the state of virus aggregation is of considerable interest to investigators studying viral infectivity, appreciable clumping of virus particles is potentially troublesome in any of the particle counting methods. The adverse effects of particle aggregation are minimized in the thin-section counting methods, since pelleted virus suspensions tend to become one uniform aggregate.

Another factor which should be considered when selecting a particle counting method is the relative ease with which virus particles can be

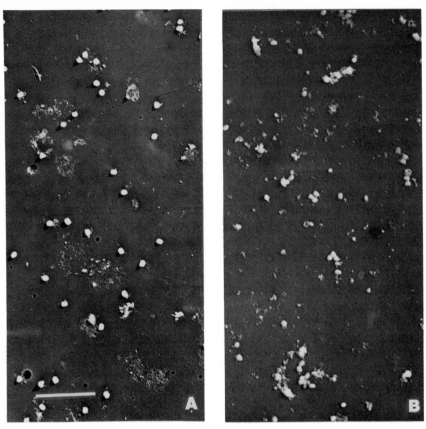

Fig. 6.21 Vaccinia virus particles prepared by the agar sedimentation procedure. (A) Well dispersed particles prepared immediately following lysis from cells. (B) Aggregated particles prepared from the same suspension as A after standing for 1 hr. The bar equals 2 μm. (Courtesy of G. Sharp.)

distinguished from nonviral material following shadowing with heavy metals, negative staining, or thin-sectioning. Since the inception of virus particle counting in the electron microscope, there has been a tendency to count only highly purified suspensions of virus particles or crude suspensions in which virus particles are so abundant that nonviral material could be effectively reduced by dilution or mild enzymatic treatment. However, it is often desirable to count impure samples containing few virus particles in relation to cellular debris, and for many viruses reliable recognition may be accomplished only in thin-sectioned samples where characteristic viral internal structures can be seen. For studies in which virus particle recognition may be troublesome, utilization of a thin-section counting procedure may be well worth the added time it takes to prepare specimens for counting.

When selecting a particle counting procedure, the electron microscopist may be interested in knowing the relative sensitivities of the various counting procedures. However, most investigators who have used the counting methods differ in their statements of sensitivity. Sharp (1965) has pointed out that part of the discrepancy lies in the investigator's willingness to count or scan large microscope fields. Sensitivity is inversely proportional to the square of the microscope magnification, so that there is considerable advantage in using the lowest magnification consistent with virus particle recognition. Relatively large portions of virus particle suspensions can effectively be sampled in such microscope fields following sedimentation. Therefore, the sedimentation counting procedures are probably more sensitive than the reference particle methods in which drops or droplets of virus particle suspensions are permitted to dry down. The thin-section counting procedures are essentially modifications of the sedimentation counting methods and, since virus particles may be more reliably recognized and counted in thin sections, it seems probable that these more recently devised counting methods are even more sensitive.

Virtually all of the described particle counting methods have their limitations, and it is not known which method yields the greatest overall accuracy. The precision of the spray-droplet procedure was reported to be $\sim \pm 10\%$ of the mean when the ratio of virus to PSL particles was unity (Luria et al., 1951). However, additional error resulting from the aggregation of PSL spheres may lower the precision to only $\sim \pm 30\%$ (Watson, 1962a). It is recommended that the investigator interested in the precision of the spray-droplet technique read the statistical analysis of this method by Breeze and Trautman (1960).

The precision of the agar sedimentation procedure was reported to be $\sim \pm 16\%$ of the mean (Sharp, 1965). However, the loss of varying amounts of viral specimens during the pseudoreplication process may alter this precision (Sharp, 1965; Müller and Nielsen, 1970).

Standard deviations ranging from 2% to 13% of mean values for counts derived by the thin-section counting method were reported (Miller et al., 1973). As in the agar sedimentation procedure, some virions may be lost when the supernatant fluid is separated from the sediment to be counted, and caution must be exercised to insure that virus particles are not dislodged from membranes during processing for thin-section electron microscopy.

Upon using one of the particle counting methods for the first time, the electron microscopist should check the precision of his own results. This

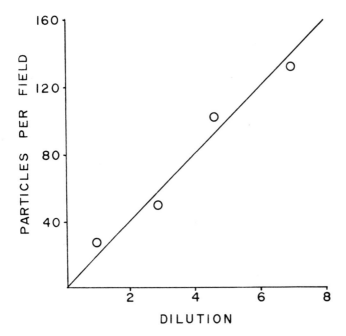

Fig. 6.22 Hypothetical regression plot showing a linear relationship between the number of virus particles counted per field and dilution of virus particle suspension.

may be accomplished by comparing his counts with counts derived by another method or by determining if a linear relationship can be obtained for several dilutions of a given virus particle suspension. Figure 6.22 shows a hypothetical dose–response regression plot which might be obtained following the above procedure.

Finally, the microscopist may wish to combine the desirable attributes of two or more of the existing particle counting methods. However, he may choose to use his own ingenuity, and devise some other method commensurate with his particular requirements and resources.

The author is grateful to Dr. M. A. Hayat for the opportunity to contribute to this volume; to Dr. Leon Dmochowski and Dr. D. Gordon Sharp for critically reviewing the manuscript; to Dr. Patton T. Allen, who critically reviewed the manuscript and whose collaboration made possible the thin-section particle counting method described in this chapter; to the scientists who kindly provided illustrations for the text; and to Mr. Robert Collins for his excellent photographic assistance. This work was supported in part by Contract NO1-CP-33304 within The Virus Cancer Program and in part by Grants RR-05511 and CA-05831 of the National Cancer Institute, National Institutes of Health, U. S. Public Health Service.

References

Backus, R. C., and Williams, R. C. (1949). Small spherical particles of exceptionally uniform size. *J. Appl. Phys.*, **20**, 224.

——— (1950). The use of spraying methods and of volatile suspending media in the preparation of specimens for electron microscopy. *J. Appl. Phys.*, **21**, 11.

Ball, F. L., and Harris, W. W. (1968). Particle counting and detection. *Proc. 26th Ann. Meet. Electron Microsc. Soc. Amer.* (Arceneaux, C. J., Ed.), p. 104. Claitor's Publishing Division, Baton Rouge, Louisiana.

——— (1972). Predictability of quantitation of small virions by electron microscopy. *Proc. Soc. Exptl. Biol. Med.*, **139**, 728.

Breese, S. S., and Trautman, R. (1960). Examination of spray-droplet particle counting as a measure of virus concentration. *Virology*, **10**, 57.

Brenner, S., and Horne, R. W. (1959). A negative staining method for high resolution electron microscopy of viruses. *Biochem. Biophys. Acta*, **34**, 103.

Crane, H. R. (1944). Direct centrifugation onto electron microscope specimen films. *Rev. Sci. Inst.*, **15**, 253.

Gehle, W. D., and Smith, K. O. (1970). Enumeration of virus particles in ultrathin sectioned pellets. *Proc. Soc. Exptl. Biol. Med.*, **135**, 490.

Geister, R., and Peters, D. (1963). Ein vereinfachtes direktes Zahlverfahren für Virus-Suspensionen ab 10^5 Partikel/ml. *Z. Naturforsch.*, **18b**, 266.

Haschemeyer, R. H., and Myers, R. J. (1972). Negative staining. *In:* Principles and Techniques of Electron Microscopy: Biological Applications, Vol. 2 (Hayat, M. A., Ed.), p. 101. Van Nostrand Reinhold Company, New York.

Hayat, M. A. (1970). Principles and Techniques of Electron Microscopy: Biological Applications, Vol. 1. Van Nostrand Reinhold Company, New York and London.

Horne, R. W., and Nagington, J. (1959). Electron microscope studies of the development and structure of poliomyelitis virus. *J. Mol. Biol.*, **1**, 333.

Isaacs, A. (1957). Particle counts and infectivity titrations for animal viruses. *Adv. Virus Res.*, **4**, 111.

Kellenberger, E., and Arber, W. (1957). Electron microscopical studies of phage multiplication. I. A method for quantitative analysis of particle suspensions. *Virology*, **3**, 245.

Luria, S. E., Williams, R. C., and Backus, R. C. (1951). Electron micrographic counts of bacteriophage particles. *J. Bacteriol.*, **61**, 179.

Mathews, J., and Buthala, D. A. (1970). Centrifugal sedimentation of virus particles for electron microscopic counting. *J. Virol.*, **5**, 598.

Miller, M. F., Allen, P. T., Bowen, J. M., Dmochowski, L., Hixson, D. C., and Williams, W. C. (1972). Particle counting of partially purified RNA tumor viruses by a thin sectioning technique. *Proc. 30th Ann. Meet. Electron*

Microsc. Soc. Amer. (Arceneaux, C. J., Ed.), p. 292. Claitor's Publishing Division, Baton Rouge, Louisiana.

Miller, M. F., Allen, P. T., and Dmochowski, L. (1973). Quantitative studies of oncornaviruses in thin sections. *J. Gen. Virol.*, **21**, 57.

Morgan, C., Rose, H. M., and Moore, D. H. (1956). Some physical and chemical properties of Coxsackie viruses A9 and A10. *J. Exptl. Med.*, **104**, 171.

Müller, G., and Nielsen, G. (1970). Elektronenmikroskopische Partikelzählung in der Virologie II. Ursachen Systematischer Fehler bei Sedimentationverfahren. *Arch. Ges. Virusforsch.*, **30**, 22.

Overman, J. R., and Tamm, I. (1956). Equivalence between vaccinia particles counted by electron microscopy and infectious units of the virus. *Proc. Soc. Exptl. Biol. Med.*, **92**, 806.

Peachey, L. D. (1958). Thin sections. I. A study of section thickness and physical distortion produced during microtomy. *J. Biophys. Biochem. Cytol.*, **4**, 233.

Pinteric, L., and Taylor, J. (1962). The lowered drop method for the preparation of specimens of partially purified virus lysates for quantitative electron micrographic analysis. *Virology*, **18**, 359.

Rhim, J. S., Smith, K. O., and Melnick, J. L. (1961). Complete and coreless forms of reovirus (ECHO 10). Ratio of number of virus particles to infective units in the one-step growth cycle. *Virology*, **15**, 428.

Riedel, G., and Ruska, H. (1941). Uebermikroskopische Bestimmung der Teilchenzahl Eines Sols über dessen aerodispersen Zustand. *Kolloid Z.*, **96**, 86.

Sharp, D. G. (1949). Enumeration of virus particles by electron micrography. *Proc. Soc. Exptl. Biol. Med.*, **70**, 54.

——— (1960). Sedimentation counting of virus particles via electron microscopy. *In: Proc. 4th Intern. Conf. Electron Microsc., Berlin, 1958.* Springer-Verlag.

——— (1963). Total multiplicity in the animal virus–cell reaction and its determination by electron microscopy. *Ergeb. Mikrobiol.*, **36**, 215.

——— (1965). Quantitative use of the electron microscope in virus research. Methods and recent results of particle counting. *Lab. Invest.*, **14**, 831.

——— (1968). Multiplicity reactivation of animal viruses. *Prog. Med. Virol.*, **10**, 64.

———, and Beard, J. W. (1952). Counts of virus particles by sedimentation on agar and electron micrography. *Proc. Soc. Exptl. Biol. Med.*, **81**, 75.

———, and Buckingham, M. J. (1956). Electron microscopic measure of virus particle dispersion in suspension. *Biochim. Biophys. Acta*, **19**, 13.

———, and Overman, J. R. (1958). Enumeration of vaccinia virus particles in crude extracts of infected tissues by electron microscopy. *Proc. Soc. Exptl. Biol. Med.*, **99**, 409.

Smith, K. O., and Benyesh-Melnick, M. (1961). Particle counting of polyoma virus. *Proc. Soc. Exptl. Biol. Med.*, **107**, 409.

Smith, K. O., and Gehle, W. D. (1969). Pelleting viruses and virus-infected cells for thin-section electron microscopy. *Proc. Soc. Exptl. Biol. Med.*, **130**, 1117.

Smith, K. O., and Melnick, J. L. (1962). Electron microscopic counting of virus particles by sedimentation on aluminized grids. *J. Immunol.*, **89**, 279.

Smith, K. O., and Sharp, D. G. (1960). Interaction of virus with cells in tissue cultures. I. Adsorption on and growth of vaccinia virus in L cells. *Virology*, **11**, 519.

Soehner, R. L., and Dmochowski, L. (1969). Induction of bone tumors in rats and hamsters with murine sarcoma virus and their cell-free transmission. *Nature*, **224**, 191.

Strohmaier, K. (1967). A new procedure for quantitative measurements of virus particles in crude preparations. *J. Virol.* **1**, 1074.

Watson, D. H. (1962a). Electron-micrographic particle counts of phosphotungstate-sprayed virus. *Biochim. Biophys. Acta,* **61**, 321.

——— (1962b). Particle counts of herpes virus in PTA negatively stained preparations. *In: Proc. 5th Intern. Conf. Electron Microsc.*, Vol. 2 (Breese, S. S., Ed.), paper X4. Academic Press, New York.

———, Russell, W. C., and Wildy, P. (1963). Electron microscopic particle counts on herpes virus using the phosphotungstate negative staining technique. *Virology*, **19**, 250.

Williams, R. C., and Backus, R. C. (1949). Macromolecular weights determined by direct particle counting. I. The weight of the bushy stunt virus particle. *J. Amer. Chem. Soc.*, **71**, 4052.

7. ULTRAMICRO-INCINERATION OF THIN-SECTIONED TISSUE

Wayne R. Hohman

Department of Obstetrics and Gynecology, Los Angeles County–USC
Medical Center, Los Angeles, California

INTRODUCTION

Ultramicroincineration combined with electron microscopy furnishes a technique which destroys the organic components of a sample but leaves the inorganic components at or near their point of origin for observation at the ultrastructural level. For biologists, this means that metals and minerals may be differentiated from their surrounding organic environment and observed within cells. Ultramicroincineration itself provides only a means of visualizing inorganic constituents of cells, but, combined with other techniques, an ultimate goal is to identify specific elements in ash residues *in situ*. Although several problems limit this goal at the present time, there is much interest in not only the roles but also the sites of action of metals and minerals in many areas of biological research. Therefore, ultramicroincineration could become a valuable aid to the electron microscopist in the study of cellular chemistry and structure.

Microincineration, as classically applied by light microscopists, uses energy from heat to remove organic material from specimens. Heat energy may also be used for incineration at the electron microscope level. In addition, an alternate method has recently been applied to biological material which uses chemical energy instead of heat to "incinerate" specimens. This method, called low-temperature ultramicroincineration, employs the energy from excited oxygen to break organic bonds and leave the inorganic substances relatively near their positions in the untreated sample. The choice between high- and low-temperature ultramicroincineration is dependent upon the goals of a specific application, but the low-temperature method is considered to yield more detailed preservation and finer resolution for use in electron microscopy.

Of many applications of incineration techniques, ultramicroincineration of thin-sectioned tissue for electron microscopic examination has a relatively recent history. Probably the first attempt to incinerate ultrathin tissue sections for the electron microscope was by Fernández-Morán (1952). He used freeze-dried thin sections (50–200 nm thick) of nerve tissue mounted on aluminum or silicon monoxide (SiO) films and incinerated them in a muffle furnace (or with the electron beam). Technical problems prevented any precise conclusions on the natural distribution of inorganic constituents, but the methods were especially innovative at that time. It is interesting to note that it was also in the fifties when Turkevich (1959) first recorded the suggestion that excited-oxygen incineration might be used for the "study of biological tissue sections," although he primarily was concerned with the role of electron stains.

Thomas (1962) revived interest in the incineration of ultrathin sections for the electron microscope by using high-temperature incineration on gray to light-gold methacrylate sections, but this work was on bacterial spores and not on tissue. A second demonstration that incineration of sections for the electron microscope was practical was performed by Desler and Pfefferkorn (1962) who used high temperatures to incinerate an artificial system consisting of silicon dioxide particles embedded in methacrylate and mounted on SiO films and platinum grids. The inorganic particles were relatively undisturbed while the methacrylate was burned away. The second attempt after Fernández-Morán to incinerate thin-sectioned tissue with high-temperature ashing was accomplished by Arnold and Sasse (1963) who briefly reported ashing sections of rat tissues (lung, pancreas, and kidney), notably exposing titanium dioxide particles from the organic material of lung tissue sections.

The first publication to seriously establish ultramicroincineration as a realistic tool for thin-sectioned material at the electron microscope level was by Thomas in 1964. Particularly significant was his application of Turkevich's suggestion to use low-temperature incineration on ultrathin sections. Although this work was a continuation of his interest in bacterial spores (Thomas, 1962), he pioneered the development of low-temperature ashing techniques that eventually were adapted for incinerating thin-sectioned tissues.

Subsequently, ultramicroincineration was used for ultrathin sections of several nontissue specimens. For instance, high- and low-temperature incineration was employed for isolated mitochondria (Thomas and Green-awalt, 1964, 1968) and Tipula iridescent virus particles (Thomas, 1965, 1969); high-temperature incineration was employed for strontium-loaded mitochondria (Greenawalt and Carafoli, 1966); and low-temperature incineration was employed for fossilized teeth (Doberenz and Wykoff, 1967).

Specifically referring to thin-sectioned tissue, high-temperature ultramicroincineration has been used for sections of rat epididymis and kidney cortex (Boothroyd, 1968) and rat intestine and bone (Martin and Matthews, 1970; Matthews *et al.*, 1970; Sampson *et al.*, 1970). After Thomas' (1964, 1969*) successful use of low-temperature incineration on bacteria, viruses, and mitochondria, Hohman (1967) and Hohman and Schraer (1967, 1972) applied low-temperature ultramicroincineration to thin-sectioned tissue. They used the hen shell gland, a calcium transporting tissue, as an example to demonstrate in detail some of the potentials and limitations of this technique when used to reveal the "mineral ultrastructure" of cells. Subsequent to the detailed publication by Hohman and Schraer (1972), the author became aware of independently performed low temperature ultramicroincineration of thin-sectioned biological materials in a dissertation by Frazier (1971). His subjects also involved calcification, specifically the early mineralization of developing bones, teeth and tendons. Frazier (1971) solved the problem of distinguishing electron opaque areas due to mineral deposits from those due to electron stains by using low temperature ashing (sometimes partial ashing or etching) in lieu of stains. Only the intrinsic mineral content of the specimens remained after low temperature ashing. An excellent review on the current status of spodography (the technique of producing ash patterns) at both high and low temperatures for both light and electron microscopes has been presented by Thomas (1969).

ULTRAMICROINCINERATION PREPARATION

General Techniques

The general techniques used for ultramicroincineration are similar to those used for conventional electron microscopy; however, some of the specific details are quite different. For example, grids are used but they cannot be of copper; support films are used but they cannot be of carbon; fixatives are used but osmium tetroxide is avoided; and stains are

---

* Through the courtesy of R. S. Thomas, the author had access to a prepublication version of this paper in 1966.

also avoided. However, if a few alternative materials and methods are substituted for those normally used, this technique is readily adaptable to any conventional electron microscope laboratory that has access to an instrument capable of high- or low-temperature incineration.

Grids

The primary consideration for selecting grids is that they must not oxidize when heated, or react with excited oxygen. Copper grids do oxidize when heated in air and catalytically react with excited oxygen. Punched (200 mesh) molybdenum grids (Hohman and Schraer, 1972) and titanium or stainless steel grids (Thomas, 1964, 1969) have been used satisfactorily for low-temperature ashing; gold, platinum, and stainless steel grids (Thomas, 1964, 1969) and nickel-plated copper grids (Boothroyd, 1968) have been used for high-temperature ashing. Gold, platinum, and nickel grids, as well as copper, are unsatisfactory for low-temperature incineration unless the power level of the LTA is maintained at a low level throughout ashing. Frazier (1971) showed that it is *possible* to use copper grids in the excited oxygen and Thomas (1974b) has used gold grids with low power levels. However, if the power rises above critical levels, copper (as well as gold, platinum, and nickel) will catalyze reactions with atomic oxygen that causes the temperature to rise dramatically and destroy the unique advantage of using low temperature ashing. Therefore, these catalytically active grids cannot be recommended, but if cost, availability, or convenience favor them, they may be used with caution.

Secondary considerations for selecting grids are hole size and method of manufacture. Because of problems in preparing support films on grids (discussed below), grids with relatively small holes (200 mesh) and flat surfaces are desirable. Both of these features provide the necessary contact area between the support films and grids. Grids 100 mesh or larger and woven wire grids generally add extra problems during preparation of the support films and during subsequent handling.

Support Films

Since both heat (in air) and excited oxygen destroy organic materials, the conventional carbon support films cannot be used for ultramicroincineration. Useful support films must be inorganic, electron-transparent, strong, stable in the electron beam, and insoluble in water. Although several alternatives exist, SiO is generally used (Williams, 1952; Hass and Meryman, 1954; Kafig, 1958).

The methods used to prepare SiO support films are similar to those for making carbon support films, but the properties of SiO films cause them to be more difficult to make and handle. One way to prepare SiO-coated grids is to evaporate SiO directly onto Formvar- or collodion-coated grids in a vacuum evaporator. However, unlike carbon films, the resulting SiO support films are not flat and appear to sag. Thomas (1964, 1969) used grids prepared in this way, burning the collodion films away with a muffle furnace or low-temperature asher either before mounting the sections or during incineration of the sections. One problem with these grids is that variations in shadowing angle cannot be avoided, but this may not be important in many studies. This method is the simplest and probably is the best starting point for a newcomer to the field.

Using Thomas' technique, Boothroyd (1968) found that fold lines resulted as the sections mounted over the sagging support films dried. He tried several alternative techniques: (1) SiO was evaporated onto Formvar-coated glass slides and the composite film floated onto water where the grids were applied. This method also did not yield flat films. (2) Layers of carbon and SiO were evaporated onto a mica surface and this composite film floated onto water where the grids were applied. This method produced the best results, but it was difficult to accomplish routinely. (3) SiO was evaporated onto the sections after they were mounted on naked grids. This method was complicated by incinerating and shadowing the underside of the grids (since the SiO was on top), and also did not result in flat films, but this was the method Boothroyd adopted.

Hohman and Schraer (1967, 1972), in their search for ways to prepare flat SiO support films, arrived independently at a method similar to Boothroyd's first method. Although this method was considered to yield improved results compared to Thomas' method, the final SiO films still sagged to some degree.

Since none of these methods produced ideal, flat, smooth, SiO support films, any one of them could be used without gaining or losing any serious advantages. One of these methods (Hohman and Schraer, 1967, 1972) will be described in detail to serve as a guide for making SiO support films. Glass slides dipped in a detergent solution and wiped dry without rinsing are coated with Formvar by dipping them in 0.5% Formvar in ethylene dichloride (Drummond, 1950). SiO is evaporated onto the Formvar-coated slide. The author put 6 mg (excess quantity) of degassed (see Hass and Meryman, 1954), powdered SiO into a tungsten boat (made from 1 mil, 1.25 in. × 0.25 in. tungsten ribbon) located 10 cm directly below a slide. The SiO was evaporated for 5 min at 25 amp filamen current in a vacuum of $\sim 4 \times 10^{-4}$ Torr.

It is important to note that if different parameters are used (e.g., a boat of different size or material), a different evaporation current and time will be necessary. The evaporation current and time were experimentally determined by heating the SiO to a point just above the minimal temperature at which evaporation occurs. The effectiveness of evaporation was estimated by observing interference colors of the SiO evaporated onto a small piece of front-surfaced mirror (a heavy chromium layer on a glass slide makes an excellent front-surfaced mirror). The colors provide an estimate of the SiO film thickness in a way similar to that used to judge section thickness in microtomy (Hass and Meryman, 1954; Kafig, 1958). A light yellow or gold color is recommended, and that is the color produced by the specific conditions just described.

After the slides are scored with a sharp tool into squares slightly larger than the grids the combined films are floated onto water, where they reflect silver under ordinary fluorescent light. The detergent residue facilitates separation of the support films from the slides. The grids are placed on the floating films (dull slide facing down) and picked up and inverted over pegs by the "ring tool and peg" method of Drummond (1950).

After drying, the films are relatively flat and smooth under low-power magnification. However, the Formvar layer is on top of the SiO and must be removed before the sections are mounted. This is done by incinerating the grids under the same conditions to be used for the sections. Unfortunately, this final step causes the slight sagging of the films over the grid holes that was mentioned above. The resulting grids have SiO films alone attached directly to the grids without intermediate layers.

Attention to detail in the preparation of the SiO support films is necessary because SiO films alone are very brittle and shatter easily; Formvar films alone are flexible and handle easily but are destroyed by the ashing. The characteristics of SiO–Formvar combinations range progressively between these two extremes as the SiO thickness is increased. Therefore, handling problems occur if the SiO is too thin or too thick. The light-yellow interference color mentioned above will yield satisfactory SiO thicknesses, estimated to be less than 50 nm. Thomas (1969) reported that his support films are ∼17.5–35.0 nm thick. Once a method is developed that is successful, repeating the process can become routine.

Thomas (1974b) has recently furnished insight into the causes of the problems encountered in preparing SiO films. He theorized that the sagging of the films is due to a conversion of SiO to SiO_2, the latter occupying a greater volume. This conversion can occur rapidly (and perhaps completely) when the SiO films are exposed to oxygen in the LTA (as in the last step of the author's procedure); or the conversion can occur slowly (and probably incompletely) when the SiO films are simply exposed to air (as during storage). Thomas (1974b) supported this theory by evaporating SiO onto a glass slide, intentionally converting the SiO to SiO_2 by heating in air (the slide providing a rigid, unsaggable support), and then removing the films and mounting them on grids. As he anticipated, the SiO_2 films were completely flat. However, this success was cancelled by the discovery that the films were so fragile that use of the grids was almost impossible. Thomas (1974b) concluded that perhaps the sagging of the SiO films is a fortuitous advantage rather than a disadvantage because this event may relieve tension from the films and thereby permit them to survive the stresses of subsequent handling. Unless a specific application requires flat films, it is recommended that one of the film preparation methods mentioned above be used in spite of their common problem with unflat surfaces (also see Frazier, 1971). The details of this progress on SiO films will be published elsewhere in the future.

Specimen Preparation

The procedures for preparing biological specimens for spodography are the same as those used for conventional electron microscopy. The choice of fixatives and buffers depends upon the goals of the project. The fixatives and buffers normally used in electron microscopy may be used in conventional ways if they are part of the desired results, or if they are considered in the interpretation of the results. However, if the research goal is to observe the intrinsic mineral–metal composition of cells or other subject matter, it is important to avoid introducing inorganic substances that would confuse interpretation of the results.

A recommended all-organic procedure is to fix in glutaraldehyde buffered with S-collidine (2,4,6-trimethylpyridine) that has been adjusted to the desired pH with acetic acid (instead of the customary HCl). Postfixation in osmium tetroxide should be eliminated. If osmolality adjustment is desired, sucrose may be used. Dehydration may be with alcohol or acetone, and embedding in any of the common embedding media should be satisfactory, although the author's experience is with Epon. The embedding medium is removed during the incineration process. For detailed embedding procedures, the reader is referred to Hayat (1970, 1972).

If osmium tetroxide fixation is used, it should be noted that the fate of osmium during ashing is not completely clear. Reduced osmium in the sample might reoxidize to OsO_4 and vaporize. On the other hand, it could remain as a heavy metal in the ash residue. Hohman and Schraer (1972) were able to observe only high-resolution differences between glutaraldehyde-fixed only and glutaraldehyde-osmium tetroxide fixed tissues.

Other comparisons have been made (Arnold and Sasse, 1963; Thomas and Greenawalt, 1968; Boothroyd, 1968; Thomas, 1969; Frazier, 1971), but there is no consensus of opinion.

There is one disadvantage to this preparative procedure that causes perhaps the greatest limitation of the ultramicroincineration technique. In every step (fixation, washing, dehydration, and even to some extent embedding) the tissues pass through solvents. While the purpose of incineration is to discover the locations of minerals and metals in the untreated sample, the solvent nature of water and other solutions must remove and rearrange some of the substances of interest. Several workers (Renaud, 1959; Boothroyd, 1964, 1968; Thomas, 1964, 1969; Hohman and Schraer 1972) have demonstrated losses of metallic substances into the solutions used for these various preparative steps. Using these standard procedures, the only reasonable approach is to confine studies to "structure-bound" inorganic substances. Some possible ways to avoid this limitation are discussed later.

Sectioning and Staining

Sectioning may be done in the conventional way for spodography, by floating sections on water, but this raises the problem discussed for specimen preparation: the solvent properties of water. Because the sections are cut thin, this step may be more critical than the other steps to this point. Boothroyd (1964) has shown that thin sections cut for electron microscopy lose water-soluble material while floating on water. As mentioned above, if water is used in the sectioning trough, the results should be interpreted as identifying structure-bound substances. In an effort to reduce soluble mineral loss, Boothroyd (1964) used a saturated $Ca_3(PO_4)_2$ solution (pH 6.1) and Thomas (1969) used a 0.001 ammonium bicarbonate solution (pH 7.5) in the sectioning trough. Specimens that contain dense mineral areas or granulas could be especially aided by these preventive procedures.

The best section thickness cannot be given here because it depends upon the reason for incineration, the nature of the specimens, and the method of incineration. In general, the thinner the section, the better the preservation of the ash. This is illustrated in Figs. 7.1–7.4 for low-temperature ashing, which show nuclei from sections that were less than 100 nm (Fig. 7.1), 500 nm (Fig. 7.2), 1,000 nm (Fig. 7.3), and 4,000 nm (Fig. 7.4) thick before ashing. The clear zone surrounding each nucleus provides an indicator for specimen distortion. The nucleus in Fig. 7.1 is well defined with a minimal clear zone around its periphery. Figure 7.2 shows a small but definite clear zone that is characteristic of 500 nm sections. Figure

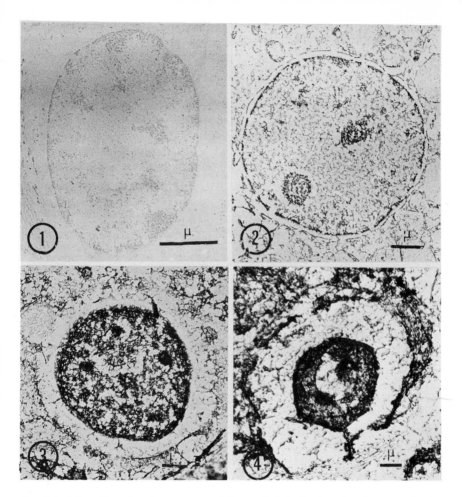

Fig. 7.1 Ash residue of a low-temperature-incinerated shell gland nucleus from an ultrathin section (glutaraldehyde-fixed, ~ 70 nm thick). Note the absence of a clear zone around the nucleus and the small amount of ash distortion. ($\times$15,400.)

Fig. 7.2 Ash residue of a low-temperature-incinerated shell gland nucleus from a 500 nm section (glutararaldehyde-fixed, osmium-postfixed). Note the narrow but distinct clear zone around the nucleus qnd the slightly disorganized appearance of the ash. ($\times$8,800.)

Fig. 7.3 Ash residue of a low-temperature-incinerated shell gland nucleus from a 1,000 nm section (glutaraldehyde-fixed, osmium-postfixed). Note the wide clear zone around the nucleus and the disorganized appearance of the ash. ($\times$6,200.)

Fig. 7.4 Ash residue of a low-temperature-incinerated shell gland nucleus from a 4,000 nm section (glutaraldehyde-fixed, osmium-postfixed). Note the very wide clear zone around the nucleus and the extreme distortion of the ash residue from this thick section. ($\times$5,600.)

7.3 shows a much wider clear zone than Fig. 7.2, and Fig. 7.4 shows a wider clear zone than Fig. 7.3.

After a little experience, the thickness of a section may be recognized by simply observing these clear zones, assuming that the quality of preparation is approximately equal. The clear zones appear to form during incineration as the supporting organic matrix is destroyed and the inorganic substances collapse upon the support films. The greater the "collapse distance" (the thicker the section), the more shrinkage and distortion will occur in the resulting ash pattern. If the specimens contain dense quantities of intrinsic mineral, they will be less sensitive to section thickness, and thicker sections would yield less distortion and be quite satisfactory. However, very thin sections have their disadvantages as well. The thinner the section, the more difficult it is to see the ash residue in the electron microscope and the smaller the quantity of ash present on the grid. Therefore, it is usually best to determine a compromise thickness for the sections that is somewhere between these extremes. For these reasons, the author eventually used 500 nm sections for most of his work.

Thomas (1964) used gray to light-gold sections, finding the thicker sections to be more useful with low-temperature ashing than at high temperatures. He noted that coalescence of the ash into droplets during high-temperature ashing prevented sections thicker than about 70 nm from being very useful. His specimens, however, were bacterial spores that had very dense inorganic cores. Boothroyd (1968) used sections ranging from silver to deep-gold (50–120 nm) for his high-temperature ashing, and he found that the thicker sections in this range were more convenient than the thinner ones. The best solution to this difficulty is to incinerate several thicknesses and pick the thickness that is most useful for the specific problem.

There is one additional point that is related to section thickness. It is essential that the sections have satisfactory contact with the support films before incineration. An easily recognized example of poor specimen contact occurs wherever there are folds in the sections. In a spodogram, these areas appear as broad, blank pathways containing unrecognizable clumps and wisps of ash. A more subtle example is shown in Fig. 7.5, which is a glutaraldehyde- and osmium-tetroxide-fixed, 500 nm thick, shadowed specimen. The cells themselves and the nuclei appear to have been well preserved, but the collagen between cells rises from the surface of the support film. In other circumstances, the residue from collagen is much better preserved, as shown in Fig. 7.14. When a section has poor contact with its support film, the collapse distance is considerable and therefore the distortion of the ash is very great.

Staining, of course, normally is avoided for ultramicroincineration stud-

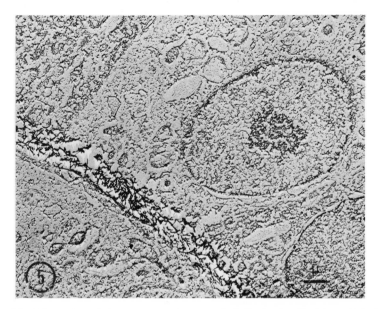

Fig. 7.5 Ash residue of low-temperature-incinerated shell gland cells (glutaraldehyde-fixed, osmium-postfixed, 500 nm section, platinum-shadowed). The ash from the nucleus and cytoplasm of these cells is well preserved, but the ash from the collagen between these cells was distorted during incineration, as shown by the shadows. (×6,200.)

ies. Stains may be used if there is a specific reason for adding a heavy metal or if the staining reaction itself is the subject. Mitochondria from a lead-stained section are shown in Fig. 7.16 as a sample of this application. Other examples may be seen in a paper by Hohman and Schraer (1972).

Previewing and Mapping

An apparent advantage of incinerating thin sections is that they may be previewed before ashing, ashed, and reviewed after ashing. In practice, this feature proves to be more disappointing than anticipated. The first problem is that unosmicated and unstained sections are difficult to observe in the electron microscope. The second is that it is time-consuming to relocate the same spot on the grid after ashing, and sometimes the specific location will have been destroyed. But even more important, the electron beam appears to affect the section in an unexplainable way so that the relatively mild low-temperature ashing conditions cannot overcome.

An example of this phenomenon is shown in Figs. 7.6 and 7.7. This par-

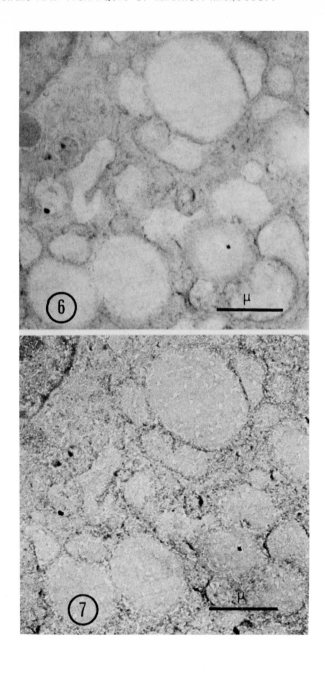

ticular section is from tissue postfixed in osmium, therefore this preview micrograph (Fig. 7.6) has slightly more contrast than a tissue fixed with glutaraldehyde alone would have. The postview micrograph (Fig. 7.7) shows a definite overlying "muddy" appearance that is not apparent in nonpreviewed spodograms (cf. Fig. 7.5). This effect which could be a contamination layer or a cross-linking by the electron beam is especially noticeable within the membrane-bound vacuoles. It is possible that ashing is not complete because of a contamination layer. Thomas (1969) also has noticed and discussed this problem. Previewing can be used, but it must be used with caution, and nonpreviewed sections should be compared with previewed sections to determine the extent of contamination.

Since it is sometimes desirable to find the same section more than once in the electron microscope, and since the ashed sections often are difficult to locate, various mapping procedures have been used (Boothroyd, 1968; Thomas, 1969; Hohman and Schraer, 1972). In general, grids are examined after the sections are mounted on them and the location of each section is recorded (by a drawing or a photograph) in reference to a fixed point. Marked grids can be used for this purpose, or a notch may be cut into the grids. Using this "map" for a guide and placing the sample in the electron microscope with the same orientation each time, any given location on a grid may be quickly located. In this way, a section may be previewed, treated, and reviewed as many times as required. If the ashed section is too faint to be observed, small artifacts or purposely applied markers (e.g., SiO_2 particles) may be used for orientation in locating a specific spot. Although mapping a grid is time consuming, it will save much time and effort whenever it is necessary to relocate specific sections.

ULTRAMICROINCINERATION

Instruments

For high-temperature ashing, either a standard laboratory muffle furnace or a vacuum evaporator can be used. Both of these instruments are com-

Fig. 7.6 Preview of a section from the shell gland that is unincinerated and unstained (glutaraldehyde-fixed, osmium-postfixed, $\sim$ 250 nm thick). Osmium and the sample's inherent density are the only sources of contrast in this somewhat thick section. ($\times$18,400.)

Fig. 7.7 Postview after low-temperature incineration of the identical location shown in Fig. 7.6 (same preparation as Fig. 7.6, but after incineration). The overlying gray layer perforated with holes was caused by contamination from the electron beam during the previewing and was not removed by the low-temperature ashing. This problem is especially visible over the large vacuoles which normally contain little or no ash after incineration. ($\times$18,400.)

monly available in most electron microscopy laboratories and can be used very simply when accurate measurement of the temperature is not required. If the temperature is to be determined accurately (as it should be for fine-structure studies), a more elaborate arrangement is required, usually consisting of a thermocouple attached near the sample. When using a muffle furnace, grids may be ashed in porcelain containers. Thomas (1964) described a thermocouple attachment for use in a muffle furnace. Thomas (1964) also used a vacuum evaporator by ashing grids on a platinum ribbon mounted between the evaporator electrodes. Boothroyd (1968) also used the terminals inside a vacuum evaporator as a source of power, but he designed his own oven by wrapping nichrome wire around a pyrex tube.

For excited-oxygen incineration, a low-temperature asher (LTA) is required. This is the only item used in ultramicroincineration that the average electron microscope laboratory may not have available. Early LTAs were made by Tracerlab (Models LTA-500 and LTA-600), but these models have been improved, expanded, and renamed by LFE Corporation (Waltham, Mass.). Their current models are the LTA-505 (five 2 in chambers), LTA-504 (four 3 in chambers), LTA-302 (two 3 in chambers), and the LTA-302-M (two 2 in chambers). Recently, an economical, compact, one-chamber LTA system has been developed by Tegal Corporation (Richmond, Calif.). Low-temperature ashers also are available from Harrick Scientific (Ossining, N.Y.), International Plasma (Hayward, Calif.) and Technics (Alexandria, Va.). Frazier (1971) described a custom-made LTA that seemed to function similar to the commercial instruments. If an LTA is to be used only for electron microscopy, the smallest and most economical models available are adequate because the samples are small and relatively few in number.

Theory

High-Temperature Incineration

The energy of heat is employed to destroy organic bonds and remove organic material from samples. This is a very efficient and effective method of incineration, but it also is a very harsh treatment. Melting, coalescing, and charring accompany burning and contribute to distorting the ash residue.

Low-Temperature Incineration

The chemical energy from excited oxygen is employed to destroy organic bonds and remove organic material from samples. The excited oxy-

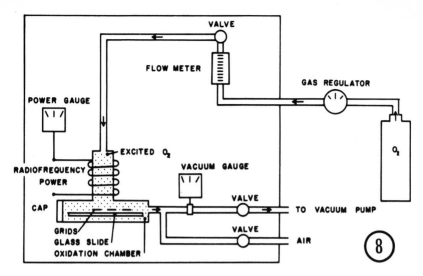

Fig. 7.8 Schematic diagram of a low-temperature asher. The arrows show the flow of oxygen when the LTA is in operation. The stippled area in the oxidation chamber shows the approximate region where the pinkish glow-discharge is visible. See text for details.

gen receives its energy from a radiofrequency electromagnetic field. A generalized schematic diagram of a low-temperature asher is illustrated in Fig. 7.8. This instrument converts relatively unreactive molecular oxygen into an excited plasma (see Gleit and Holland, 1962; Gleit, 1963, 1965, 1966; Hollahan, 1966). The oxygen is obtained from standard-grade commercial oxygen bottles. Some impurities such as water vapor and nitrogen are desirable; therefore, highly purified oxygen is not recommended.

The molecular oxygen flows from the oxygen bottle through valves and regulators to the chamber where the radiofrequency electromagnetic field is applied. The excited plasma formed by the electromagnetic field is a complex mixture of atomic oxygen, molecular oxygen in excited states, ionized oxygen, free electrons, and various impurities and reaction products (Foner and Hudson, 1956; Herron and Schiff, 1958; Elias *et al.,* 1959; Ogryzlo and Schiff, 1959; Kaufman and Kelso, 1960; Marsh and coworkers, 1963, 1965a, 1965b; Gleit *et al.,* 1963; Bersin, 1965; Hollahan, 1966). The excited plasma, visible as a pinkish discharge (stippled area in Fig. 7.8), passes over the grid and exits through a vacuum pump that maintains the entire process in a mild vacuum. The action of the excited oxygen might be described as plucking the organic molecules out of the sample, leaving the inorganic residue behind to collapse upon its support. For this reason, low-temperature incineration is considered to be relatively gentle compared to high-temperature incineration.

Comparison of High- and Low-Temperature Incineration.

There are two main factors to compare between high- and low-temperature incineration: (1) the quantity of substances retained and (2) the amount of distortion of the ash pattern caused by the incineration procedure. Comparative studies indicate that high-temperature ashing vaporizes more material than low temperature ashing, and that the harshness of high-temperature ashing is more likely to displace the inorganic substances than the relative gentleness of low-temperature ashing. Before supporting these statements, it should be noted that the importance of these factors depends upon the specific application of ultramicroincineration, i.e., the specific elements of interest and the resolution desired. For some purposes, either high- or low-temperature ashing will answer the questions; for other purposes, one or the other will be necessary.

The question of the amounts of substances retained is a quantitative one and can therefore be substantiated in a straightforward manner. Known substances can be measured before or after performing high- and low-temperature incineration. The following examples have been selected from the recovery data by Gleit and Holland (1962).

Radioactive sodium (as NaCl plus blood) was 100% retained during low-temperature ashing and 100% lost during high-temperature ashing (24 hr at 400°C). On the other hand, radioactive manganese (as $MnCl_2$ plus blood) was 100% retained during low- and high-temperature ashing under the same conditions used for Na. Therefore, if Mn is the element of interest, high- or low-temperature ashing would be adequate; if Na is the element of interest, low-temperature ashing would be essential. The retention of Na during low-temperature incineration is particularly significant because both Na and K are considered to be volatile during conventional ashing at 550–600°C. Both Na and K were quantitatively retained (98.2–99.0%) from rat liver microsomes after low-temperature ashing at 100–125°C (H. Sanui, cited in Hollahan, 1966).

Low-temperature incineration always yields the same or improved elemental retention compared to high-temperature ashing. The author does not know of a single example where the high-temperature ashing yield is greater than the low-temperature ashing yield. One point of caution in interpreting recovery data is that the chemical form of an element influences its retention. For example, iodine as NaI was 30% retained in the low-temperature asher, but iodine as $NaIO_3$ was 100% retained. Both forms were lost in the high-temperature asher (Gleit and Holland, 1962).

For recovery data on other elements using high- and low-temperature incineration, the reader is referred to Thiers (1957), Gorsuch (1959),

Pijck *et al.*, (1961), Gleit and Holland (1962), Gleit (1⌐
lahan (1966), Hamilton *et al.*, (1967), and Thomas (1⌐

The question of the extent of distortion of the ash pa⌐
cineration is less a quantitative one than the question ⌐
and therefore not as simple to substantiate. Thomas (1964) ⌐
high- and low-temperature incineration and concluded that with low-t⌐
perature incineration the ash does not melt, but rather aggregates into a
fine-textured granular reticulum, allowing much sharper definition of the
ash patterns. Thomas and Greenawalt (1968) compared high-temperature (500°C and 600°C) and low-temperature incineration of mitochondria. High-temperature ashing retained mitochondrial granules (calcium
phosphate) but the mitochondrial membranes disappeared. Low-temperature ashing also retained the granules, in addition to distinct ash residues
from the mitochondrial membranes.

The choice of high or low temperature must be governed by the degree
of resolution required. If the subjects of interest are relatively large, distinct, inorganic objects such as granules, high- or low-temperature ashing
may be used; but if the subjects of interest are the mineral content of
fragile, predominantly organic objects (such as mitochondrial membranes)
low-temperature ashing is essential. Hohman and Schraer (1972) compared their results on only low-temperature incineration of tissue sections
with those of Boothroyd (1968) on only high-temperature incineration of
tissue sections and concluded that the low-temperature ash patterns were
richer and more detailed, especially in cytoplasm (different tissues were
compared, however). An additional discussion on this topic has been presented by Thomas (1969). Since the low-temperature ash patterns appear
to contain more ash, is the organic material completely removed? Thomas
and Greenawalt (1968) low-temperature-ashed mitochondria and then
reincinerated them at 600°C in a muffle furnace. The doubly incinerated
specimens appeared the same as the specimens which were only low-
temperature-ashed indicating that the high-temperature procedure caused
no additional ashing of the specimens.

If maximal resolution and mineral retention is required, low-temperature ashing should be used. If optimal conditions are not required, either
method may be used. For some applications, high-temperature ashing
may be preferred. For example, previewing seems to have no effect on the
final ash pattern when high-temperature ashing is used (Thomas, 1969),
whereas low-temperature ashing is not sufficiently strong to remove the
contamination layer from the electron beam, as shown in Figs. 7.6 and 7.7.

Operation

High-Temperature Incineration

Although there are many opinions on the temperature and duration for optimal incineration, high-temperature ashing is usually done at 550–660°C (sometimes 400–500°C for longer periods of time) for 10–30 min.

Low-Temperature Incineration

Conditions for operating an LTA vary according to the instrument and the material to be incinerated. The author employs the minimal conditions that will maintain a discharge. Using a Tracerlab LTA-600, the author uses a low power level (50 watts) at minimal reflected power, 70 cc/min oxygen flow rate, and ~0.7 mm Hg pressure. Ashing time is 20 min, although this is excessive; repeat incineration causes no additional change in the specimen. Since thin sections are very small, it is reasonable that mild ashing conditions should be effective.

Although heat is not applied during low-temperature ashing, the reactions of the excited oxygen and specimen are exothermic and some heat is generated. Using high power levels and bulk materials, the temperature can go over 100°C, but it is assumed that the low power levels and small size of the electron microscope specimens keep the temperature well below 100°C (see Hohman and Schraer, 1972, for a more detailed discussion of sample temperature).

Postincineration

Sometimes the ashed sections and support films are destroyed by the electron beam after low-temperature incineration. If this problem occurs, a light layer of carbon evaporated on top and bottom of the grid solves this problem (invert the grid over a hole to protect its surface film). Sandwiching the sample between light carbon layers seems more effective than relying on one heavy layer. The ashed grids require no special handling. The ash adheres to the grids well, does not absorb water, and seems to survive all the handling described.

Shadowing with platinum or another heavy metal may be used to reveal the three-dimensional features of the ash. The contrast of many ashed specimens is sufficient without shadowing. However, shadowing can enhance faint ash patterns, makes the ash patterns more distinct (Fig. 7.5), and sometimes provides additional information on the ash that would have been missed without shadowing (Fig. 7.13). The shadowing material does not contribute to the reinforcement problem solved with carbon.

APPLICATIONS

Present

The spodograms presented in this section have been selected to demonstrate the capabilities of low-temperature incineration. Figures 7.9–7.16 are low temperature incinerated thin sections of a tissue, the hen shell

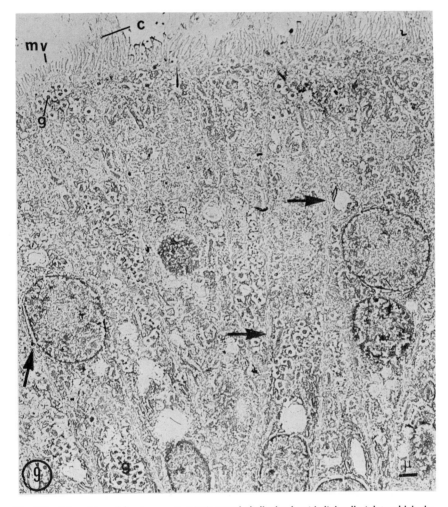

Fig. 7.9 Ash residue of low-temperature-incinerated shell gland epithelial cells (glutaraldehyde-fixed, osmium-postfixed, ∼ 500 nm section). Note the ash residues from cell membranes (arrows) which occur only in osmium-fixed tissue. c, cilia; mv, microvilli; g, granules. (From Hohman and Schraer, 1972). (×4,400.)

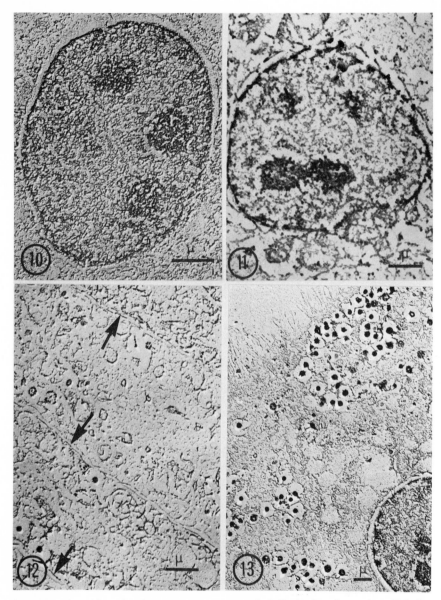

Fig. 7.10 Ash residue of a low-temperature-incinerated shell gland nucleus (glutaraldehyde-fixed, 500 nm thick section, platinum-shadowed). ($\times$11,200.)

Fig. 7.11 Ash residue of a low-temperature-incinerated shell gland nucleus (glutaraldehyde-fixed, osmium-postfixed, 500 nm thick section). ($\times$9,500.)

gland. These examples are intended to serve as a guide for realistically deciding whether or not a potential problem could be aided by ultramicroincineration.

Figure 7.9 is a low magnification of a spodogram that shows the general distribution of ash within several sectioned epithelial cells. This micrograph is from a section that was approximately 500 nm thick before ashing; the tissue was fixed with glutaraldehyde followed by osmium tetroxide. Note the obvious retention of ash in the nuclei, and also the generous ash retention in the cytoplasm. The nuclei are partially surrounded by clear zones, although they are not as distinct as in Fig. 7.2 (also see Figs. 7.10, 7.11, and 7.13). Small dense granules can be recognized in the cytoplasm, and empty vacuoles are apparent, but the rest of the ash in the cytoplasm is disorganized.

If more detail is desired for a specific problem, thinner sections would be required. To illustrate this point: mitochondria cannot be recognized readily in Fig. 7.9, but they can be preserved in thin sections, as can be seen in Fig. 7.14. The ash from microvilli and cilia are visible along the free border of the cells (top of the electron micrograph). The detail obtainable from this spodogram is seen in the visible cell membranes (arrows) between the cells. This feature is characteristic of low-temperature-ashed, glutaraldehyde-fixed, osmium-postfixed tissue. The ash from the cell membranes is not apparent in spodograms from only glutaraldehyde-fixed tissue. This feature is significant because most of the cellular structures appear to be similar with or without osmium.

Figures 7.10 and 7.11 are detailed views of low-temperature-ashed nuclei from 500 nm sections. Fig. 7.10 was glutaraldehyde-fixed and shadowed with platinum; Fig. 7.11 was glutaraldehyde-fixed and osmium-postfixed but not shadowed. Both spodograms contain the clear zones around the nuclei which are characteristic of 500 nm sections. The clear zones also appear to a lesser extent around the nucleoli, which are very clear in both figures. The surrounding cytoplasmic ash is relatively undefined. Note that the same features are present with (Fig. 7.11) and without (Fig. 7.10) osmium postfixation. Shadowing produces a crisper

---

Fig. 7.12 Ash residue of low-temperature-incinerated shell gland epithelial cells (glutaraldehyde-fixed, osmium-postfixed, 500 nm section, platinum-shadowed). Note the ash residue from cell membranes (arrows). ($\times$9,500.)

Fig. 7.13 Ash residue of low-temperature-incinerated shell gland epithelial cells containing dense granules (glutaraldehyde-fixed, 500 nm section, platinum-shadowed). The ash from the granules formed dense deposits as they contracted during ashing. Note the clear zones around the nucleus and nucleoli. ($\times$5,000.)

appearing electron micrograph (Fig. 7-10), but it does not provide additional information in this case.

Figure 7.12 is a detailed view of the ash from cellular membranes. The membranes separating four cells stand out very clearly (arrows) in this shadowed spodogram from a 500 nm section. As mentioned for Fig. 7.9, this is characteristic of tissues postfixed with osmium. Whether the ash residue in the membrane is osmium or was formed by the action of osmium is not known.

Figure 7.13 was chosen because it demonstrates the type of problem that is ideally suited for application of ultramicroincineration. Electron microscopy of hen shell gland epithelial cells reveals many dense granules near their luminal borders. Since the shell gland is a calcium transporting tissue, the mineral content of these granules is important for elucidating the role of the granules and the function of these cells. Low-temperature-ashing a 500 nm section from the tissue fixed with glutaraldehyde only yielded the results shown in Fig. 7.13.

The ash from each granule formed a dense ball in the center of a clear space. Apparently the granules contracted during ashing by a mechanism similar to that for nuclei and nucleoli. In stained, conventional electron micrographs, in unstained, unashed electron micrographs, and in thin-sectioned, ashed electron micrographs, the granules occupy the entire area defined by the clear spaces. Note also that the contracted ash from the granules has significant three-dimensional shape, rather than being a flat, low-profile circles. The membranes between cells are not visible, since this is an all-organic-prepared specimen.

Figures 7.14–7.16 show low temperature ashed mitochondria prepared by three different methods. The mitochondria ash residue in Fig. 7.14 is from a very thin section (less than 100 nm) of a tissue fixed with glutaraldehyde followed by osmium tetroxide. The mitochondrial membranes and cristae are minimally distorted by the low-temperature ashing process. This spodogram shows not only the fine detail that is obtainable when the sections are thin enough, but is also an example of a spodogram that is almost invisible in the electron microscope. Also, note the excellent preservation of the collagen in this section, which contrasts with the poor preservation in Fig. 7.5.

Figure 7.15 was prepared in the same way as Fig. 7.14 except that it was sectioned at 500 nm and platinum shadowed. The increased distortion of the ash is observed in both the mitochondria and the nearby cytoplasm. More commonly, mitochondria are not recognizable in the ash from 500 nm sections (Fig. 7.9).

Figure 7.16 shows osmium-fixed mitochondria from an ashed thin section that had been stained with lead citrate before ashing. The lead appears to have formed small granules during the ashing process, but the mitochondrial outline has remained visible.

In addition to the examples shown here, other applications of low- and high-temperature incineration to sections from tissues and other substances may be obtained from the references mentioned earlier. For those whose interests lie in areas other than thin-sectioned biological materials, Thomas (1974a) has recently collected a large number of references (over 300) to applications of low-temperature plasmas which includes fields such as mineralogy, metallurgy, polymer chemistry, biomedical sciences, and air pollution; instruments such as the scanning electron microscope, light microscope, electron diffractometer, and electron microprobe; techniques such as etching, concentrating substances, dispersing substances, preparing thin films, testing the organic nature of substances; and preparing replicas; and specimens such as plant leaves, cigars, rice hulls, seashells, wool fibers, wood, coal, teeth, bone, polymers, cotton, etc., in addition to the applications to thin-sectioned, biological tissues using low temperature incineration and transmission electron microscopy mentioned in this chapter.

Future

Minimizing or eliminating the limitations of a technique is always one of the hopes for the future in any experimental field. Perhaps the most severe limitation of ultramicroincineration of thin-sectioned material is the solvent nature of the preparative solutions. Most of the work has been necessarily limited to structure-bound metallic–mineral substances. These solvent steps may be eliminated by sectioning frozen, freeze-dried, or freeze-substituted samples. Fernández-Morán (1952) and Arnold and Sasse (1963) tried to use freeze-dried material without great success; however the technique has been advancing steadily. Cryosectioning in general, without special reference to incineration, has been developing as a technique (Nei, 1974; Simard, 1975). One possible way to avoid floating the thin sections on water is to cut them dry. This is difficult to do, but could be very worthwhile for some specimens. Elimination of the aqueous steps of the procedures would provide a more accurate ash distribution and would especially aid studies on water-soluble metals.

A second valuable advance in ultramicroincineration could occur if a technique for identifying the composition of the ash would be developed. X-ray microanalysis is an apparent solution to this problem (Gleit, 1966), but there are two factors that have prevented much progress on ashed, thin-sectioned tissue: the quantity of ash and the resolution of the X-ray microanalysis instruments.

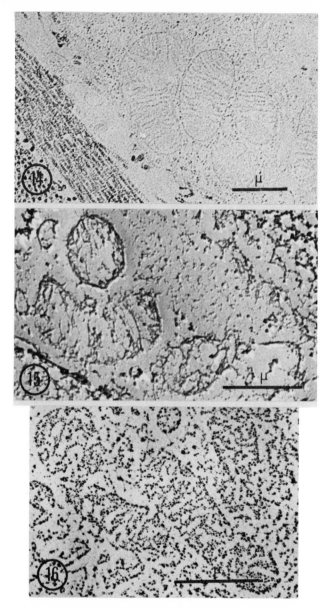

Fig. 7.14 Ash residue of low-temperature-incinerated shell gland mitochondria from a thin section (glutaraldehyde-fixed, osmium-postfixed, 100 nm thick section). Note the detailed preservation of the mitochondrial membranes and cristae, and also the segmented collagen. (From Hohman and Schraer, 1972.) (×14,000.)

Figure 7.17 shows the result of a preliminary attempt to measure the ash composition of a 500 nm thick section in an electron microprobe. The figure is an electron micrograph that shows the dense trails which formed when the focused spot of a microprobe had previously traversed the section. The trails provide a precise record of the pathway of the beam and this can be correlated with the X-ray readout on the elements. Calcium and phosphorous were shown to be present in the ash of Fig. 7.17, because the counts for Ca and P from the tissue were higher than those from the Epon alone (see the transition zone near the top of the micrograph). However, it was not possible to distinguish the locations of these elements within the cellular substructure. Thomas (1969) had some limited success with X-ray microanalysis on ashed virus crystals. Low temperature incineration combined with transmission electron microscopy provides both quantitative and qualitative information on the location of minerals in the specimen. The addition of X-ray microanalysis adds quantitative information on the composition of the ash. When combined, these three techniques yield more information than any one alone. Combined scanning electron microscope and X-ray microanalysis instruments are becoming popular, so that progress in this area could be forthcoming.

There are many other potential methods to increase the usefulness of ultramicroincineration. Some histochemical techniques may be adaptable to the electron microscopy of ash residues. For some purposes, inorganic substances may be added to biological tissues. For example, ashing has been used to show the results of strontium (Greenawalt and Carafoli, 1966) and calcium phosphate (Thomas and Greenawalt, 1968) uptake by mitochondria. Electron stains may be studied with the aid of incineration. Tissue pathogenicity caused by metal imbalances would be an interesting subject for incineration studies. Tissues exposed to industrial pollutants could prove to be interesting.

In some cases it may be significant that a structure disappears during ashing. The manner in which a structure ashes may be informative. For example, the granules in Fig. 7.13 would have disappeared if they were entirely organic, or they would have remained unchanged by the ashing if they were inorganic. The fact that they contracted into dense balls

---

Fig. 7.15 Ash residue of low-temperature-incinerated shell gland mitochondria from a thicker section than that of Fig. 7.14 (glutaraldehyde-fixed, osmium-postfixed, 500 nm section, platinum-shadowed). The mitochondrial ash is much more distorted in this thicker section than in Fig. 7.14. (From Hohman and Schraer, 1972.) ($\times$21,800.)

Fig. 7.16 Ash residue of low-temperature-incinerated shell gland mitochondria that were stained with lead citrate before ashing (osmium-fixed, $\sim$100 nm section, lead-citrate stained). ($\times$24,000.)

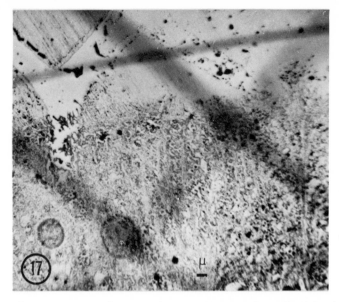

Fig. 7.17 Electron micrograph illustrating the contamination tracks produced by the beam of the electron microprobe as it traversed the section (glutaraldehyde-fixed, osmium-postfixed, 500 nm section). This section was low-temperature-ashed, scanned in the electron microprobe, and then examined and photographed in the electron microscope. (From Hohman and Schraer, 1972.) ($\times$2,900.)

provides a clue that they may be organometallic complexes, the inorganic residue remaining after the supporting organic framework was destroyed. Incineration may be used to thin certain types of specimens. Thomas (1974a) lists etching as an application for oxygen plasmas. Partial incineration could reveal structural information. Since metals and minerals are important in many fields, there are many possible applications of ultramicroincineration in the future.

CONCLUSIONS

Ultramicroincineration has proven to be capable of becoming a useful technique in the electron microscope laboratory. Tissues have been shown to contain significant quantities of inorganic substances distributed throughout all parts of their cells. It is possible to achieve excellent preservation of the ash from thin-sectioned tissues. Resolution can be at the level of biological membranes. Research goals can be kept within the limitations of the technique, or they can be directed toward reducing these limitations. Although the technique can be developed into an elaborate procedure, it can also be used in most electron microscope laboratories without much added equipment or expense.

Ultramicroincineration is a method which alters thin sections in order to gain information regarding tissues. In this sense, ultramicroincineration shares a common feature with many techniques employed by electron microscopists. Staining, shadowing, fixation, dehydration, embedding, cutting, freezing, freeze-drying, freeze-substitution, freeze-fracturing, critical-point drying, coating, etc., alter tissues to make it possible to gain information about them. Ultramicroincineration, like these techniques, has advantages, disadvantages, and optimal conditions for its use. It seems that the technique of ultramicroincineration has advanced sufficiently to be added to the list of techniques available to electron microscopists.

The author wishes to thank Harold Schraer of Pennsylvania State University for many encouraging discussions and for the use of the facilities of his electron microscope laboratory, where most of the author's work on low-temperature ultramicroincineration has been accomplished. Acknowledgment is also due to Mr. Gene Marcus for the photography, and Miss Ann Malkie for typing the manuscript.

References

Arnold, M., and Sasse, D. (1963). Das Aschebild im Elektronenmikroskop. *Acta Histochem.*, **3** (Suppl.), 204.

Bersin, R. (1965). LTA-600 low temperature dry asher, technical applications guide. Tracerlab, Inc., Richmond, California.

Boothroyd, B. (1964). The problem of demineralization in thin sections of fully calcified bone. *J. Cell. Biol.*, **20**, 105.

——— (1968). The adaptation of the technique of micro-incineration to electron microscopy. *J. Roy. Microsc. Soc.*, **88**, 529.

Desler, H., and Pfefferkorn, G. (1962). Die Veraschung als Präparationmethode in der Elektronenmikroskopie. *In:* Electron Microscopy (Breese, S. S., Ed.) **1**, EE-1. Academic Press, Inc., New York.

Doberenz, A. R., and Wyckoff, R. W. G. (1967). The microstructure of fossil teeth. *J. Ultrastruct. Res.*, **18**, 166.

Drummond, D. G. (1950). The practice of electron microscopy. *J. Roy. Microsc. Soc.*, **70**, 1.

Elias, L., Ogryzlo, E. A., and Schiff, H. I. (1959). The study of electrically discharged O_2 by means of an isothermal calorimetric detector. *Can. J. Chem.*, **37**, 1680.

Fernández-Morán, H. (1952). The submicroscopic organization of vertebrate nerve fibers. An electron microscope study of myelinated and unmyelinated nerve fibres. *Exptl. Cell Res.*, **3**, 282.

Foner, S. N., and Hudson, R. L. (1956). Metastable oxygen molecules produced by electrical discharges. *J. Chem. Phys.*, **25**, 601.

Frazier, P. D. (1971). Electron Microscopic Investigation of Mineralizing Tissues. Dissertation for Ph.D., University of Washington, Seattle, Washington. Available from University Microfilms, Inc., Ann Arbor, Michigan.

Gleit, C. E. (1963). Electronic apparatus for ashing biologic specimens. *Amer. J. Med. Electron.*, **2**, 112.

———— (1965). High frequency electrodeless discharge system for ashing organic matter. *Anal. Chem.*, **37**, 314.

———— (1966). Recovery and chemical analysis of submicrogram particles. *Microchem. J.*, **10**, 7.

Gleit, C. E., and Holland, W. D. (1962). Use of electrically excited oxygen for the low temperature decomposition of organic substances. *Anal. Chem.*, **34**, 1454.

————, Holland, W. D., and Wrigley, R. C. (1963). Reaction kinetics of the atomic oxygengraphite system. *Nature* (London), **200**, 69.

Gorsuch, T. T. (1959). Radiochemical investigations on the recovery for analysis of trace elements in organic and biological materials. *Analyst*, (London), **84**, 135.

Greenawalt, J. W. and Carafoli, E. (1966). Electron microscope studies on the active accumulation of Sr^{++} by rat-liver mitochondria. *J. Cell Biol.*, **29**, 37.

Hamilton, E. I., Minski, M. J., and Cleary, J. J. (1967). The loss of elements during the decomposition of biological materials with special reference to arsenic, sodium, strontium and zinc. *Analyst*, **92**, 257.

Hass, G., and Meryman, H. T. (1954). Silicon monoxide and its use in electron microscopy. *12th Ann. Mtg. Electron Microsc. Soc. Amer.*, Highland Park, Illinois.

Hayat, M. A. (1970). Principles and Techniques of Electron Microscopy: Biological Applications, Vol. 1. Van Nostrand Reinhold Company, New York and London.

———— (1972). Basic Electron Microscopy Techniques. Van Nostrand Reinhold Company, New York and London.

Herron, J. T., and Schiff, H. I. (1958). A mass spectrometric study of normal oxygen and oxygen subjected to electrical discharge. *Can. J. Chem.*, **36**, 1159.

Hohman, W. (1967). A study of low temperature ultramicroincineration of the avian shell gland mucosa by electron microscopy. Dissertation, The Pennsylvania State University, University Park, Pennsylvania.

———— and Schraer, H. (1967). Ultramicroincineration studies of the avian shell gland. *J. Cell Biol.*, **35** (2, Pt. 2), 57A.

———— and Schraer, H. (1972). Low temperature ultramicroincineration of thin-sectioned tissue. *J. Cell Biol.*, **55**, 328.

Hollahan, J. R. (1966). Analytical applications of electrodelessly discharged gases. *J. Chem. Educ.*, **43**, A401.

Kafig, E. (1958). Preparation of large intact unsupported evaporated films. Res. Rept. NM710100.07.01., Naval Med. Res. Inst., Bethesda, Md., **16**, 823.

Kaufman, F., and Kelso, J. R. (1960). Catalytic effects in the dissociation of oxygen in microwave discharges. *J. Chem. Phys.*, **32**, 301.

Marsh, H., O'Hair, T. E., Reed, R. and Wynne-Jones, W. F. K. (1963). Reaction of atomic oxygen with carbon. *Nature* (London), **198**, 1195.

Marsh, H., O'Hair, T. E., and Wynne-Jones, W. F. K. (1965a). Oxidation of carbons and graphites by atomic oxygen, kinetic studies. *Trans. Faraday Soc.*, **61**, 274.

Marsh, H., O'Hair, T. E., and Reed, R. (1965b). Oxidation of carbons and graphites by atomic oxygen, an electron microscope study of surface changes. *Trans. Faraday Soc.*, **61**, 285.

Martin, J. H., and Matthews, J. L. (1970). Mitochondrial granules in chondrocytes, osteoblasts, and osteocytes; an ultrastructural and microincineration study. *Clin. Orthoped. Related Res.*, **68**, 273.

Matthews, J. L., Martin, J. H., Sampson, H. W., Kunin, A. S., and Roan, J. H. (1970). Mitochondrial granules in normal and rachitic rat epiphysis. *Calcif. Tissue Res.*, **5**, 91.

Nei, T. (1974). Cryotechniques. *In: Principles and Techniques of Scanning Electron Microscopy: Biological Applications*, Vol. 1 (Hayat, M. A. Ed.). Van Nostrand Reinhold Company, New York and London.

Ogryzlo, E. A., and Schiff, H. I. (1959). The reaction of oxygen atoms with NO. *Can. J. Chem.*, **37**, 1690.

Pijck, J., Gillis, J., and Hoste, J. (1961). La détermination du Cu, Cr, Zn, et Co dans le serum par radioactivation. *Int. J. Appl. Radiat. Isot.*, **10**, 149.

Renaud, S. (1959). Superiority of alcoholic over aqueous fixation in the histochemical detection of calcium. *Stain Technol.*, **34**, 267.

Sampson, H. W., Matthews, J. L., Martin, J. H., and Kunin, A. S. (1970). An electron microscope localization of calcium in the small intestine of normal, rachitic, and vitamin-D-treated rats. *Calcif. Tissue Res.*, **5**, 305.

Simard, R. (1975). Cryoultramicrotomy. *In: Principles and Techniques of Electron Microscopy: Biological Applications*, Vol. 6 (Hayat, M. A. ed.). Van Nostrand Reinhold Company, New York and London.

Thiers, R. E. (1957). Contamination in trace element analysis and its control. *In:* Methods of Biochemical Analysis (Glick, V. D., Ed.), p. 223. Interscience Publishers Inc., New York.

Thomas, R. S. (1962). Demonstration of structure-bound mineral constituents in thin-sectioned bacterial spores by ultramicroincineration. *In:* Electron Microscopy (Breese, S. S., Ed.), **2**, RR-11. Academic Press Inc., New York.

——— (1964). Ultrastructural localization of mineral matter in bacterial spores by microincineration. *J. Cell Biol.*, **23**, 113.

——— (1965). Ultrastructural localization of mineral constituents by microincineration and electron microscopy. *J. Cell Biol.*, **27** (2, Pt. 2), 106A.

——— (1969). Microincineration for electron-microscopic localization of biological minerals. *In:* Advances in Optical and Electron Microscopy (Barer, R., and Cosslett, V. E., Eds.), **3**, 99. Academic Press, New York.

——— (1973). Microscopical Applications of Low Temperature Plasmas. A bulletin available on request from TEGAL Corporation, Richmond, California.

——— (1974a). Use of chemically reactive, gaseous plasmas in preparation of specimens for microscopy. *In:* Techniques and Application of Plasma Chemistry (Hollahan, J. R., and Bell, A., Eds.). Wiley–Interscience, New York. In press.

Thomas, R. S. (1974b). Personal communication.

———, and Greenawalt, J. W. (1964). Microincineration of calcium phosphate-loaded mitochondria. *J. Appl. Physiol.*, **35**, 3083.

———, and Greenawalt, J. W. (1968). Microincineration, electron microscopy, and electron diffraction of calcium phosphate-loaded mitochondria. *J. Cell Biol.*, **39**, 55.

Turkevich, J. (1959). The world of fine particles. *Amer. Scient.*, **47**, 97.

Williams, R. C. (1952). High resolution electron microscopy of the tobacco mosaic virus. *Biochim. Biophys. Acta*, **8**, 227.

8. PREPARATORY METHODS FOR ELECTRON PROBE ANALYSIS

James R. Coleman and A. Raymond Terepka

Department of Radiation Biology and Biophysics, School of Medicine
and Dentistry, University of Rochester, Rochester, New York

INTRODUCTION

Current electron probe X-ray microanalyzers are based on Castaing's (1951) work. In the original instruments a focused electron beam was visually positioned on a specimen and the resulting X-rays were analyzed according to wavelength and intensity. Later instrumentation added the capability of operating in the scanning mode (Cosslett and Duncumb, 1956), and it became feasible to utilize nonoptical imaging so that an operator could work with images produced by the interaction of the electron beam with the sample. Thus, it became possible to compare sample current, backscattered electron, and secondary electron images, which convey morphological information, with X-ray images of the same specimen. This greatly facilitated the correlation of specific chemical composition with morphological features.

These capabilities were extremely useful in fields of geology, metallurgy, and other materials sciences. However, the spatial resolution and minimum detectability limits of these early instruments limited their usefulness in biological investigations. Since then, both spatial resolution and detectability limits have been improved to the point that several areas of

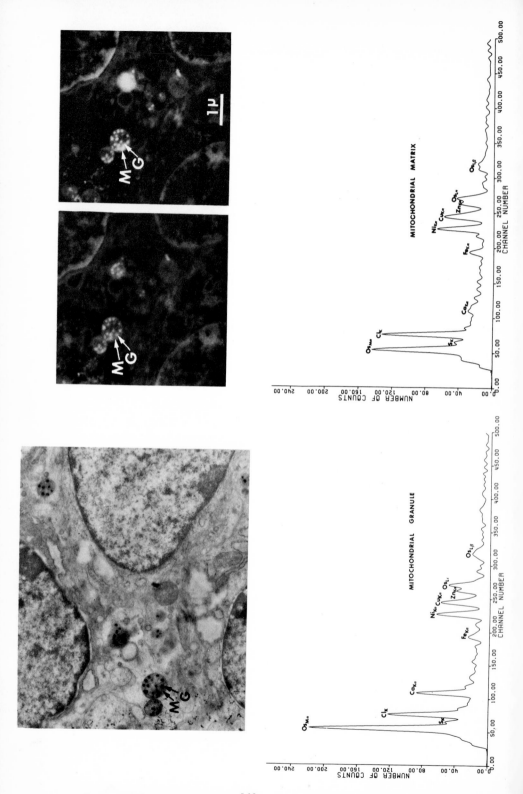

biological research are likely to benefit from the unique capabilities of electron probe analysis. For example, reports have appeared which claim to have detected elements present in concentrations as little as 0.01% by weight or in masses as small as 10^{-18} g, with spatial resolutions approaching 30 nm (Russ, 1971). Sutfin et al. (1971) have reported the analysis of single intramitochondrial granules in thin sections (Fig. 8.1).

In view of these possibilities it is surprising that electron probe microanalysis has not been more widely used in biological studies. The cost of the instrument ($50,000 to $100,000) is one reason that not every histochemistry laboratory is so equipped. The other, and probably more important, reason for the paucity of biological electron probe analyses is the lack of suitable methods to prepare cells and tissues for analysis. Electron probe analysis detects elemental composition and is thus best suited for inorganic analyses. However, presently available methods of tissue preparation have generally aimed at preserving macromolecules. Only recently has any considerable effort been expended to develop tissue preparation methods that would preserve the distribution of small molecules, and of these, most are designed for organic molecules (see, e.g., various articles in Roth and Stumpf, 1969). As a result there is a distinct scarcity of methods designed to preserve the normal in vivo distribution of inorganic molecules.

The reasons for this scarcity may be better appreciated when one considers the nature of biological samples and the conditions under which electron probe analysis must be performed. With the accelerating voltages commonly employed, the mean free path of the electrons in the exciting

Fig. 8.1 The micrograph on the far left, which was taken in a conventional electron microscope, shows several mitochondria with dense granules. The cells are odontoblasts from the pulp of a developing rat molar. The tissue was fixed in paraformaldehyde and glutaraldehyde in a cacodylate buffer followed by osmium tetroxide. Gold-colored sections of the tissue were mounted on 75 mesh nickel grids and no staining was performed. The two micrographs on the right are dark-field scanning transmission electron micrographs of the same cells. That on the left was taken before high spatial resolution microanalysis was performed and that on the right was taken afterwards. In all three micrographs, the granule selected and analyzed is marked G and the region in the adjacent matrix is marked M. In the right-hand dark-field micrographs, the granule and matrix are obscured by electron-beam-induced contamination. That these spots are only slightly larger than the granules indicates that the volume analyzed is not much greater than that of the granule. The spots have a center-to-center distance of 100–150 nm. Hence, the spectra which are shown below originated from regions quite close to each other. The spectrum from the granule has a much greater calcium $K\alpha$ peak than that of the matrix. The osmium $M\alpha$ peak is greater in the granule while the chlorine K peak is less. The constancy of the iron $K\alpha$, nickel $K\alpha$ and copper $K\alpha$ peaks indicates that the amount of "stray" radiation coming from the specimen grid and stage is constant for the two analyses. The $M\alpha$ peak of osmium overlaps that of phosphorus, but the latter element has been identified in similar mitochondrial granules prepared without osmium fixation. The mitochondrial granules in this case represent accumulations of calcium (and probably phosphorus) and similar results were obtained when the analysis was repeated. (Courtesy of L. V. Sutfin.)

beam is too short to permit operation in air. Analysis must be performed in a vacuum. As in electron microscopy, this requires that the samples be dry or, alternatively, frozen, before entering the vacuum in order to avoid disruptive boiling that might occur upon sudden exposure to reduced pressure.

Removal of water before exposure to the vacuum of the instrument poses several problems. Many of the elements of greatest interest are distributed through the water phase of the sample (e.g., sodium and potassium), and removing the water is very likely to alter the distribution of these elements. If the change in distribution is sufficiently large, then the electron probe analysis will no longer reveal the true *in vivo* distribution of the elements of interest. Additionally, the fact that the spatial resolution and sensitivity of the electron probe are almost unique, means that there are few if any independent methods available which can be used to corroborate electron probe results. Thus the electron probe user is usually dependent on the testimony of indirect methods to evaluate the effects of any preparatory method that he may employ.

Finally, the densities of tissues (especially soft tissues) including bone and tooth, are low enough that the electron beam penetrates rather more deeply into the tissue than it would into minerals and metallic samples, as can be seen in Fig. 8.2. Since the entire volume excited by the electron beam can emit X-rays, this effectively reduces spatial resolution. In order to circumvent this consequence of low density, tissues may be cut into sections thin enough that their thickness is less than the depth of electron beam penetration. This reduces the volume of tissue emitting X-rays and improves spatial resolution (Hall, 1971). However, there is a point of diminishing returns. Since the concentration of most inorganic elements in noncalcified tissues is relatively low, as sections are cut thinner, the mass of an element in the volume excited may diminish to less than the detectability limits.

Another difficulty associated with sectioning is worth noting. Most tissues must be embedded in a supporting matrix to provide sufficient mechanical support that the tissue will not be deformed or destroyed by the relatively severe forces encountered during sectioning (Wachtel *et al.*, 1966). All successful embedding procedures depend on replacing most or all of tissue water with the embedding material. Removing water, as previously mentioned, is liable to redistribute elements dissolved in the water phase. Even the mechanics of sectioning and section handling can redistribute tissue components (Boyde and Switsur, 1963; Boothroyd, 1964; Wachtel, *et al.*, 1966; Coleman and Terepka, 1972a).

In spite of these obstacles, a growing number of investigators are employing electron probe analysis in biological studies. The purpose of this

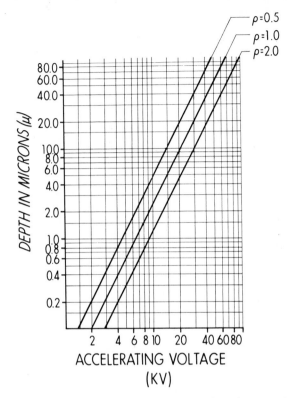

Fig. 8.2 Depth of electron beam penetration as a function of accelerating voltage and density. Calculated from the results of Andersen and Hasler (1966) and based on the work of Warner (1972).

chapter is to describe some of the techniques that have been employed to prepare samples for electron probe analysis and to suggest criteria that will help the reader decide whether these and other techniques are suitable for his purposes, and likely to produce valid analyses.

It is, however, beyond the scope of this chapter to evaluate or even merely catalog all preparative methods employed so far in electron probe analysis of histological material. This limitation results from the fact that many reports of electron probe analysis give only cursory descriptions of the preparatory techniques employed. Omission of such detail is often the result of the analysis being presented as a brief note, or similar abbreviated publication, in which space limitations preclude describing techniques in sufficient detail to permit their repetition in other laboratories. Another reason that authors may not detail and evaluate sample preparation procedures is the difficulty of testing a procedure when only indirect methods of corroborating results are available. In a way, this last point is a reflection of the almost unique capabilities of the electron probe.

Since there are so few techniques capable of the resolution and sensitivity of electron probe analysis, one is forced into the uncomfortable logical position of using electron probe analyses to justify other electron probe analyses, with each analysis depending on some sort of prior sample preparation. It hardly seems wise to disregard the potentialities of the electron probe because of this difficulty, but any investigator must live with the possibility that results may have to be reinterpreted in the light of new findings regarding a particular preparatory procedure. Thus the emphasis in this chapter will be on describing methods and providing references so that the reader may examine them first hand. Because of the reasons presented above, attempts at evaluating the techniques must be limited to somewhat general considerations, with the burden of appraising the techniques in the light of experimental findings left to those who choose to use them.

No attempt will be made to describe the range of instrumentation that is currently available, for this topic will be dealt with in detail by Weavers in Volume 5 of this treatise. For methods of quantitative analysis, the reader is referred to Andersen (1967), Beaman and Isasi (1970), Birks (1971), Colby (1968), Hall, et al. (1972), Heinrich (1968), Ingram, et al. (1972), Lehrer and Berkely (1972), Warner (1972), and Warner and Coleman (1972a, 1973).

APPLICATIONS

Electron probe analysis has been applied in two different ways to biological problems. First, it has been used as a microchemical method for the analysis of very small volumes (e.g., hundreds of picoliters) of biological fluids (Ingram and Hogben, 1968; Lechène, et al., 1969; Morel, et al., 1969; de Rouffignac, et al., 1969). This technique is employed for analyzing fluids that were in contact with just a few cells.

It has been successfully used to determine the composition of forming urine in various parts of the renal tubules of the kidney. In the kidney studies, calibrated micropipettes under a light microscope were utilized to withdraw small samples of urine from the renal tubules. The urine is placed in small uniform drops on a beryllium block and freeze-dried. The drops are then analyzed quantitatively, and usually, automatically. Lechène and co-workers (Lechène et al., 1969; Morel et al., 1969; de Rouffignac et al., 1969) have shown that this technique can be accurate, precise, and quite sensitive. Many of the difficulties encountered in the analysis of cells and tissues are avoided in this technique. Depending on the dexterity of investigators and their willingness to undertake micromanipulation, this procedure should be more widely used in the future.

The other and more common use of the probe has been as a histochemical tool used to relate elemental composition to morphological features. Thus, the following discussion will be largely restricted to this aspect of its use.

Histochemical applications of electron probe analysis may concern all, some, or only one of the elements present. The elements may exist in one or more of the following states: (1) amorphous or crystalline precipitates (e.g., calcium in bone); (2) covalently bound to a macromolecule or other organic molecule (e.g., the sulfur of cysteine residues in a protein); (3) bound by ionic or other noncovalent bond to a macromolecule or organic molecule (e.g., iron in heme); (4) free, soluble and diffusible (e.g., the potassium ion free in the cytoplasm or cytosol). The purpose of the investigation may be qualitative, to decide whether particular elements occupy certain locations; semiquantitative, to measure the relative concentration of elements at different sites; or quantitative, to measure the actual concentration of elements at various sites. If a biological specimen to be examined is a hard, mineralized tissue, it must be polished or thinned for analysis. Soft tissues must be sectioned prior to analysis, while single, thin cells or isolated cell organelles may be analyzed whole.

Each combination of these characteristics is likely to impose its own set of requirements on a preparatory procedure. A technique designed to permit the analysis of the distribution of a single element in hard tissue is not likely to be suitable for the quantitative analysis of several elements in soft tissues. It is often an advantage to examine the requirements of any particular investigation so that a preparatory technique which meets the minimum requirements may be employed, and an unnecessarily complicated one that far exceeds the needs of the investigation may be avoided. Furthermore, since the validity of the analysis, as with any histochemical technique, is strongly influenced by the preparatory procedures which precede the actual analysis, it is desirable to consider the preparatory procedure in relation to the purpose of the investigation.

CRITERIA FOR SPECIMEN PREPARATION

In general, there are four aspects of any preparative procedure that should be taken into consideration (Coleman and Terepka, 1972a). First, the morphology or normal structural relationships of the sample must be preserved. This is obvious since the purpose of the analysis is to relate elemental composition to the structural elements of the tissue or cell. However, it is not so obvious where the limits of preservation for any investigation should be set. Since the composition of various phases of cell cytoplasm is maintained by membranes 10 nm or less in thickness, morphological preservation at the electron microscope level of resolution is appropriate for analysis of cell compartments and organelles. Even in cases where one proposes only to distinguish the "inside" from the "outside" of a cell, the cell membrane must remain intact. Thus, the effect of disrupting phase boundaries within cells, or between cells, must be considered, and the likely effect on the results of any particular analysis must be evaluated.

Second, the amount of material lost or gained by the entire sample during preparation must be known. Usually this can be assessed by bulk chemical analysis of one half of the sample prior to preparation and the other half after preparation. It is rarely feasible to perform a total elemental analysis of something as complex as biological samples which may contain 20 or more elements. Usually such before and after analyses are restricted to just the few elements of interest. If the amount of loss is known, it may be compensated for in any quantitative procedures. If the amount lost is a major portion of the total present, then the concentration of the element may fall below the minimum detectable limit. The concentration of sodium in an erythrocyte is estimated at 10–20 mM; small losses could easily make the remainder undetectable.

Third, and more important than general loss or gain, is selective loss or gain. The elements of a biological system are often sequestered in compartments with quite different compositions and characteristics. It is quite possible to remove one loosely bound pool of an element, while retaining others. Thus, even if only a small, selective loss occurred, it might eliminate the contents of one pool, while others remained unchanged. Consequently, the distribution of the element elucidated with the probe would not represent the distribution that existed *in vivo*.

When it is feasible to label only some of the compartments with an isotope, it is possible to test for this occurrence. For example, some tissues have a calcium pool that is rapidly labeled upon exposure to ^{45}Ca, while other pools are far slower in exchanging their ^{40}Ca contents for the radioactive isotope. Thus, one can expose the sample to the isotope for a short duration to label the rapidly labeled pool, then divide the sample in half, analyze one-half directly for ^{40}Ca and ^{45}Ca, treat the other half according to the preparatory procedure, and analyze the treated half for ^{40}Ca and ^{45}Ca. Comparison of the specific activities of each half will show whether only one pool was changed or proportional changes occurred in each pool. Table 8.1 presents the results of such an analysis of selective calcium loss.

The final criterion concerns redistribution or translocation. The whole

Table 8.1   Comparison of Specific Activities (^{45}Ca cpm $\times$ 10^{-3}/μg Ca) in Treated and Untreated Chick Chorioallantoic Membranes

Experiment No.	Specific activity		Ratio
	Control	Treated	
1	7.2	8.9	0.81
2	7.3	7.1	1.03
3	8.5	7.9	1.08
4	6.8	5.9	1.15
5	6.9	8.2	0.84
6	6.7	6.3	1.06
7	7.6	7.9	0.96
8	7.3	8.1	0.90

sample may not lose any appreciable amount of the elements of interest but the elements may no longer occupy the sites they occupied *in vivo*. This is also the most difficult artifact to assess. No one direct method is available to detect redistribution, but, some general criteria may be helpful.

The occurrence of unusual crystals or precipitates may indicate translocation (Fig. 8.3). The formation of some crystals (e.g., calcium phosphate) can almost completely deplete the surrounding medium of the elements involved. For example, self seeding of calcium phosphate solutions occurs at a concentration higher than the solubility product of calcium phosphate (Fig. 8.4). This means that once precipitation begins, the concentration of calcium and phosphate will fall to that determined by the solubility product of the precipitate (Neuman and Neuman, 1958).

The distribution of an element of known concentration may be helpful. In many organisms, for example, the concentration of sodium is about ten times higher outside than inside the cells, and the concentration of potassium is about ten times higher inside than outside. A sharp boundary between these two phases in the probe would indicate that they had not diffused during preparation. A diffusion of one into the other would be indicted by a gradual slope in their X-ray profiles (Ingram and Hogben, 1968; Coleman, *et al.*, 1972).

It is also possible to test preparative procedures with models. One gelatin block containing the element of interest (e.g., sodium) can be placed next to another containing none of the element of interest but containing a marker (e.g., sulfur) covalently bound to the gelatin. The two joined

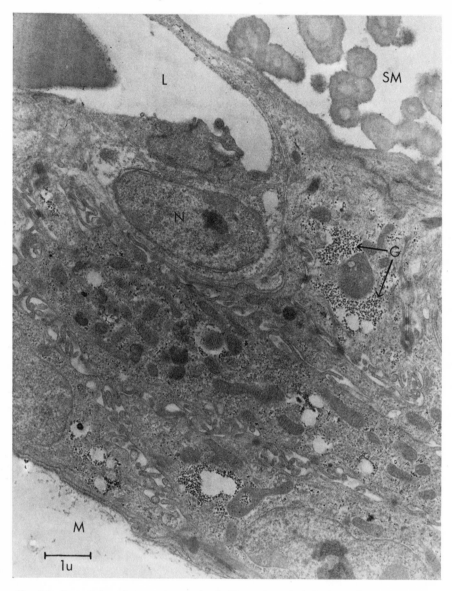

Fig. 8.3 Transmission electron micrograph of thin section of chick chorioallantoic membrane fixed in 6% acrolein in 0.1M phosphate buffer, pH 7.2, followed by postfixation with 1% OsO_4 in the same buffer. Aggregations of small granules (G) appear throughout the cell. The selected area electron diffraction patterns from these aggregates are similar to the pattern produced by calcium hydroxyapatite standards. N, nucleus; L, lumen of capillary; SM, shell membrane; M, mesoderm.

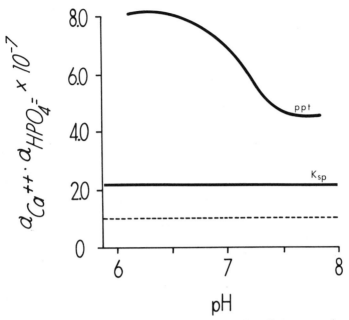

Fig. 8.4 Plot of precipitation point (ppt) and solubility product (K_{sp}), expressed as activity product of calcium and phosphate, as a function of pH. The dotted line represents the activity product of calcium and phosphorus for human serum. (Adapted from Neuman and Neuman (1958).)

blocks can be processed and X-ray profiles will indicate whether the sodium has diffused into the sulfur-containing block. The sulfur being bound to a macromolecule which is crosslinked to others through the process of gelation would not be expected to diffuse. Translocation may also be indicated if different preparative procedures (e.g., precipitation and chemical fixation vs. freeze-drying) produce different results. Finally, in unusual cases, it may be possible to look for changes in environment or association when dealing with paramagnetic metals through the use of paramagnetic resonance methods.

GENERAL PREPARATORY TECHNIQUES

In general, specimen preparation procedures so far employed in histological electron probe analyses have come under one of four categories:

1. *Conventional tissue preparation methods* similar to those used for light and electron microscopy. For soft tissues this usually involves chemi-

cal fixation with organic aldehydes and/or heavy metals (e.g., osmium), dehydration with alcohol, and embedding in plastic or paraffin, followed by sectioning. Hard tissues such as bone, tooth, and shell may be fixed with a chemical fixative, or dried, then embedded in a plastic matrix and polished with a graded series of abrasives.

2. *Precipitation methods*, which involve the addition of some exogenous material to bind the elements of interest in an insoluble complex that will withstand extraction during subsequent procedures, which usually consist of some adaptation of conventional fixation, dehydration, and embedding followed by sectioning or polishing, if necessary.

3. *Freezing methods*, which involve a first step of freezing the tissue. Subsequently the sample may be "freeze-dried" (i.e., dried in air under vacuum), "freeze-substituted" (i.e., dehydrated by replacing water with another, usually organic, solvent), or "frozen-sectioned" (i.e., sectioned in the frozen state and then dried by any of a variety of methods).

4. *Drying methods*, which involve the vaporization of water with no prior freezing or fixation. The techniques of air drying and "heat fixation" come under this last category, as well as "critical-point drying," in which the water content of the sample is replaced first by a solvent and then by a liquefied gas which sublimes directly to the gas phase when its critical temperature is exceeded (Anderson, 1965; Hayat and Zirkin, 1973; Cohen 1974). This latter technique has the advantage of drying samples without permitting them to pass through the distorting surface tension forces found at air–water interfaces.

Most tissue preparation techniques involve one or more common steps. In view of the specimen characteristics and analytical requirements outlined above, it is informative to consider, in general terms, problems that may arise and hazards which may be encountered in using preparative techniques.

Fixation

Fixation is a process which supposedly stops changes in the biological material, including changes due to self-inflicted autolysis. It also implies cessation of motion in the sample at the level of cells, organelles and molecules (see discussion in Hayat, 1970). It is obvious that one cannot fix biological material without changing it, but it is not always obvious how much change can be tolerated without interfering with the purposes of an investigation.

In dealing with elements bound covalently to macromolecules, the conventional methods of tissue preparation utilized for electron microscopy and developed to retain macromolecules are probably adequate. However, for elements not covalently bound to macromolecules, or bound to organic molecules by ionic or other bonds, one must consider the dissociation constant of the materials in the relatively large volumes of the various solvents to be encountered. Factors such as the pH dependence of binding and changes in the configuration of macromolecules as they encounter solvents of different ionic strengths and compositions should be taken into account.

However, sufficient information to predict these effects is not available and they are not readily determined. Thus, for elements not covalently bound to macromolecules, the use of precipitation techniques may be advantageous. Examples of these techniques are the use of oxalate to precipitate calcium (Fig. 8.5) (Constantin et al., 1965), pyroantimonate to precipitate mono- and divalent ions (Komnick, 1962, 1969), and lead to precipitate phosphates (Goldfisher and Moskal, 1966). It is necessary that such methods produce a very insoluble product and that the precipitant be added in sufficient concentration that the element of interest will encounter it before diffusing any appreciable distance. This means that the precipitant must be able to enter cells.

The stoichiometry of the reaction should also be known. Thus, if it is necessary to precipitate magnesium and calcium with oxalate, it is important to know that only a small portion of the total magnesium may be precipitated, while the remainder will probably be lost during subsequent exposures to aqueous solvents. Therefore, different amounts in the product finally analyzed may not mean different amounts were present in vivo. Finally, the effect of the precipitation agent on morphology must be assessed. For example, removal of divalent ions from cell membranes through the use of a precipitant may permit the membranes to become more flexible or fragile, and the introduction of excess phosphate may distort cells through the formation of aggregates of phosphate crystals (Fig. 8.3) (Coleman and Terepka, 1972a). The use of precipitation methods seems best suited for the analysis of one or a few elements. It should be noted that quantitation procedures may be adversely affected if in preserving only some elements, others are lost.

Several workers have employed freezing techniques to avoid some of these problems, the purpose being to prevent loss or redistribution by trapping cell materials in a rapidly formed matrix of frozen water. The samples may be examined while in a frozen state using a freezing stage, models of which already exist (Miller and Wittry, 1966; Bacaner et al., 1972), or freeze-dried either as intact tissue (Ingram and Hogben, 1968)

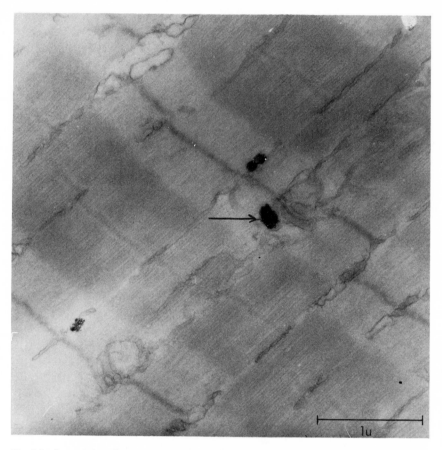

Fig. 8.5 Transmission electron micrograph of frog muscle fiber showing calcium oxalate deposits (arrow) in sarcoplasmic reticulum. The sarcolemma of the fiber was removed, the fiber was perfused with $CaCl_2$ (10 mM) and sodium citrate (80 mM) followed by sodium oxalate (10 mM) and KCl (140 mM). Fixation was in 6.5% glutaraldehyde in 0.2 M cacodylate buffer, pH 7.2, washed in buffer alone, and fixed again in 2% osmium tetroxide, with each solution containing 10 mM oxalate. Tissue was dehydrated in ethanol and embedded in Araldite for this section. No staining was carried out because this step tended to remove the oxalate deposits. (Micrograph courtesy of Clara Franzini-Armstrong.)

or as sections (Lehrer, 1971). Since damage observed in the tissue that had been frozen was attributed to the thawing process, these techniques eliminated this hazard. However, recent investigations have shown that the process of freezing itself produces significant changes in the distribution of materials in solution through the formation of eutectic phases (see Wolstenholme and O'Connor, 1969).

Aqueous solutions usually do not freeze homogeneously; instead, pure water tends to freeze out first, excluding solute from the forming ice matrix. Thus, during freezing, there will be some frozen and some still liquid phases in the solution. The phases that are still liquid will contain elevated concentrations of solute, and since the freezing point of the phase will be determined by the concentration of solute, the higher the concentration of solute in the phase, the lower the temperature (or eutectic point) at which the phase will freeze.

The formation of such eutectic phases in glycerol–water solutions was documented by Staehelin and Bertaud (1971). They used freeze-cleaving and freeze-etching techniques so that they could examine the material in the frozen state, and thus avoid any ambiguity that might occur if the solutions suffered later thawing or manipulation. Figure 8.6 shows a

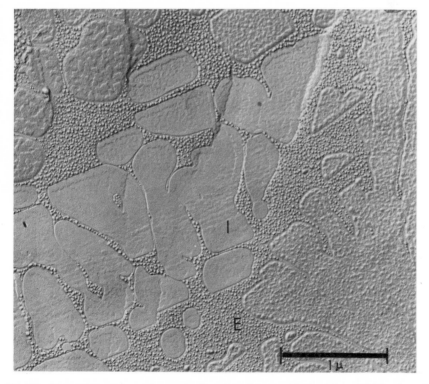

Fig. 8.6 Transmission electron micrograph of replica of freeze-etched solution of 20% glycerol in water, frozen at −150°C and replicated at −120°C. There are at least two distinct phases present which thought to be ice (I) and a glycerol hypoeutectic phase (E). The particles on the surface are formed from the condensation of water vapor prior to replication. (Micrograph courtesy of L. Andrew Staehelin.)

freeze-etched preparation of a rapidly frozen glycerol solution. The variations in appearance indicate that the water in different phases has sublimed at various rates because of differences among the glycerol–water ratios in the different phases.

Freeze-cleavage techniques were also employed (Schmitt *et al.*, 1970; reviewed by Zingsheim, 1972) to examine the effects of freezing on the distribution of macromolecules and particles. Solutions of ferritin, of albumen, of dextran, and of polyvinylpyrrolidone, and suspensions of polystyrene latex particles were shock-frozen in Freon cooled by means of liquid nitrogen, and examined by freeze-cleave and freeze-etch techniques. They found that solute molecules had been concentrated in various ways during the freezing process, and even polystyrene latex particles as large as 0.1 μm in diameter were redistributed and concentrated at the edges of grain boundaries by the growth of ice crystals.

Bank and Mazur (1973) used similar freeze-fracture techniques to examine the effects of freezing on yeast cells. Their observations led them to conclude that even at very rapid cooling rates intracellular ice phases form. They further concluded that slow freezing, which preserves the viability of yeast cells, does so because cell water is removed from the cell during extracellular ice formation, reducing ice crystal formation and resultant damage within the cells. This finding was consistent with earlier findings of Rebhun (1972), who found that exposing clam eggs to concentrated salt solutions before freezing prevented intracellular ice crystal formation by removing cell water.

In an effort to document some of the effects of freezing on erythrocytes, Farrant and Woolgar (1972) exposed cells to solutions of osmolarities they could encounter during freezing. They noted that any phases still liquid at $-3.7°C$ would be greater than 2,000 mOsM, and at this osmolarity erythrocytes would gain sodium, lose water, and begin to leak potassium. Thus any technique involving freezing, a process formerly thought benign, deserves careful scrutiny (see discussion in Rebhun, 1972). In this regard it is worthwhile noting that Lehrer and Berkeley (1972) and Ingram *et al.* (1972) have found conditions which can produce relatively uniform distributions of electrolytes in frozen dried or frozen sectioned protein solutions. It is likely that the spatial redistributions induced by freezing can be kept below the spatial resolution of electron probe analysis, and thus have minimal influence on results. However, Lehrer and Berkeley (1972) have also shown that severe redistribution artifacts can also occur, even in freeze-dried frozen sections of gelatin solutions.

Dehydration

This is accomplished by one of two general methods: gradual substitution of another material for water, or drying with no replacement of the water that is removed. Conventional dehydration procedures for the embedding materials used in light and electron microscopy prescribe gradual dehydration through a series of ethanol solutions of increasing concentration, until the sample is permeated with pure ethanol. The effects of this process are not well described, but it is known that molecules as large as glycogen can be translocated within cells during this type of dehydration. Pearse (1961) has shown micrographs of liver tissue in which glycogen has been swept to one side of cells as a result of ethanol dehydration. This manifestation of dehydration artifacts is familiar enough to histochemists to have received the name "Alkoholflucht."

Presumably, if molecules as large as glycogen, the polymers of which can have molecular weights of millions (Mahler and Cordes, 1971), can be redistributed, then other smaller molecules can also be swept through the cell during dehydration, and may stop at cell structures that form barriers to their passing. Such barriers might be cell membranes or other organelles. The result of this movement would be an apparent and artifactual concentration at the barriers that impede their movement.

Pearse (1961) suggests that suspicious patterns of distribution should be investigated by comparing the picture seen with conventionally dehydrated samples to that seen with material dehydrated in another manner, such as freeze-drying. However, even with freeze-drying there is reason to believe that some elements are progressively concentrated at the sublimation front during drying, giving an erroneous impression of distribution. Extracelluar chlorine, for example, is often found to coat cell membranes in freeze-dried preparations (Ingram and Hogben, 1968). It is doubtful that this is its real distribution; it is more likely that chlorine was adsorbed onto the first diffusion barrier it encountered during drying.

Freeze-substitution, in which frozen tissue is bathed in cold fluid until the water has been replaced by the fluid, may eliminate some of the problems associated with the sublimation fronts that occur during freeze-drying. The fluids commonly employed in freeze-substitution are organic solvents such as liquid propane, acetone, and dimethyl formamide. Thus, they are capable of dissolving many cell constituents, especially lipids and hydrophobic macromolecules. Such losses can be monitored by labeling tissue with a general label such as ^{14}C-acetate which can be incorporated into several types of metabolic pools, and then analyzing the used dehydrating fluids for radioactivity. Since many cations are bound by lipids, and since many membranes which act as phase boundaries in cells derive their permeability properties and structural integrity from lipids, lipid extraction may have significant effects on the distribution of elements in the sample as it is finally analyzed.

Frozen sectioning deserves mention under this heading. With this technique tissue is frozen at about liquid nitrogen temperatures, and sectioned directly using the ice formed within the tissue as a support to resist the deforming forces of sectioning. Depending on the freezing microtome used, sections sufficiently thin to be used for transmission electron microscopy can be cut (Bernhard and Leduc, 1967; Christensen, 1971; Appleton, 1971; Hodson and Marshall, 1971a). The thicker sections may be floated on a bath of liquid nitrogen or some similar fluid before being transferred to a substrate. Sections are then dried, either in room temperature air or at low temperatures, as in freeze-drying. Thus, the same factors that affect tissue during freezing, sectioning, and freeze-drying operate in these techniques, albeit on smaller scales.

Anderson (1967) showed that simple air drying of *Amphiuma* red blood cells could be used to preserve several intracellular elements. Rapid drying of single cells using a flame and a conducting substrate has been used with reasonable success (Coleman *et al.*, 1972; 1973a and b). Air drying or rapid drying with heat may be effective because of the fact that most biological membranes are readily permeable to water. Water may thus evaporate from cells so rapidly that no detectable redistribution occurs. The geometry of the situation may be an advantage also. Large surface area and relatively small volume promote rapid evaporation. Furthermore, during the evaporation process movement of material in a direction parallel to the shortest axis of the specimen and perpendicular to the plane of the substrate will not be detectable in these specimens because the elements would still remain within the volume excited by the electron beam.

Embedding

The penetration of tissue by an embedding medium is usually considered a benign part of sample preparation and for the most part this is true. However, as with dehydration by substitution, one must consider the likely effects of advancing fronts of embedding material on the redistribution or adsorption of elements on membranes and other structural elements. The gross defects of embedding, such as shrinking, swelling, or "explosion" within tissue, may cause redistribution, but these effects are usually signaled by their distortion of morphology (Hayat, 1970; Rebhun, 1972; Ingram, *et al.*, 1972). The temperatures at which embedding occurs will also have an effect on such processes as diffusion. In some cases, elements have been found to diffuse from tissue into epoxy embedments. An example of this is the slow dissociation of indium from nucleic acid embedded in Araldite and its diffusion into the surrounding polymerized embedding medium (Watson and Aldridge, 1961). This also points to the importance of considering the dissociation constants of elements in the various environments to which they will be exposed.

Sectioning

As mentioned previously, biological samples do not usually occur in a form that permits optimum resolution or analysis of the interior. Sectioning is frequently necessary. Some quantitative procedures require or work best with thin sections. This is a relative term usually describing a sample whose density and thickness permit partial transmission of the beam. The physical forces of shear, compression, tension and local heating acting upon tissues being sectioned have been described (Wachtel et al., 1966; Hayat, 1970). Dense, hard materials are often relocated within softer tissues during cutting (Boyde and Switsur, 1963). The larger and harder the material, or the greater the difference in density or hardness between an inclusion and its matrix, the more troublesome this is likely to be.

In addition, sections are usually floated on a liquid surface to permit surface tension forces to restore them to original dimensions after being compressed during cutting. The interaction between this fluid and the section may be of significance. If the fluid is water or an aqueous solution, then the volume of water is usually large relative to the small amount of material in a section only a micrometer or so in thickness (Boothroyd, 1964; Coleman and Terepka, 1972). Thus, even elements of low solubility are likely to be removed from the section by the medium. There have been some attempts to develop "waterproof" embedding compounds by including various silicone fluids. These may prove quite useful (Stirling and Kinter, 1967).

Examination

Biological polymers, the structural components of cells, are susceptible to radiation and heat damage. The polymers used as embedding materials also undergo various degradations when exposed to an electron beam (Bahr et al., 1965; Stenn and Bahr, 1970a and b; Hall, 1971). Certain bonds are more readily damaged than others; thus, there is differential damage throughout a specimen depending on its composition. Some polymers will crosslink and become more stable, while others will break down and may become volatile. Selective volatilization can lead to apparent changes in concentration.

In the electron probe analysis of a series of organic calcium salts, for instance, it was found that some substances consistently had higher calcium concentrations than would be expected from stoichiometry. It appeared that the material was decomposing under the influence of the beam. This was confirmed by exposing the substances to varying temperatures for different periods of time. The samples exhibited differential mass losses (Fig. 8.7). Thus, the heating of the sample induced by the electron beam was probably sufficient to incinerate some of the organic matter, and leave behind the less volatile calcium, effectively increasing the calcium concentration.

Some elements (e.g., halogens and mercury) are normally volatile. Other losses may occur as a migration phenomenon such as that described by Hodson and Marshall (1971b), or may be caused by energy transfer from the electron beam (R. Castaing, personal communication). In the latter case, enough energy is delivered to various atoms to permit them to escape from the sample. Techniques such as weighing samples before and after analysis and interference microscopy to detect mass changes can be useful in checking for radiation and heat damage. Elec-

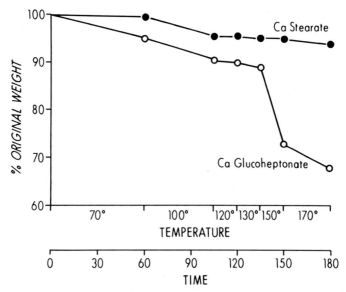

Fig. 8.7 Differential mass loss in two organic calcium salts as a function of temperature. Pellets of the salts were placed in platinum crucibles and exposed to temperatures from 70°C to 170°C for the periods noted on the abscissa. (From Warner (1972.))

tron microscope examination of replicas of samples exposed to the probe will show certain types of damage. As in transmission electron microscopy, exposure of the sample to low doses of electrons seems to stabilize the sample and prevent or minimize visible beam damage. Adequate conduction between sample and substrate is necessary to minimize heat damage as well as sample charging. Thin (1–2 μm) sections mounted on silicon or graphite disks, or quartz coated with evaporated aluminum, frequently need no other conductive coating. Extrapolation of counts to zero time and repeated scans over a sample can give some indication of whether or not damage due to beam exposure is likely.

SPECIFIC PREPARATORY TECHNIQUES

To this point, the discussion has concerned specimen preparation procedures in rather general terms, the purpose being to identify sources of possible artifacts that should be considered before employing any particular specimen preparation procedure. The previous discussion may have contained enough caveats to suggest that adequate specimen preparation is such an impossible goal that actual probe analysis may never be performed. Such is not the case, for a substantial amount of useful information has already been gained using methods that were recognized to be limited, but adequate for the analysis at hand. In this section a representative selection of preparatory methods is presented. They have been chosen to illustrate the wide variety of specimens already analyzed, and the range of preparative procedures that have been developed to meet specific needs.

Particulate Materials

Galle and coworkers (Galle, 1964a and b, 1967a and b; Galle and Morel-Maroger, 1965; Galle et al., 1968; Stuve and Galle, 1970) have analyzed several insoluble materials such as lead, gold, and calcium salts that accumulate in various tissues (see Fig. 8.8). The insoluble nature of these precipitates made it feasible to use conventional electron microscope techniques, including fixation or postfixation with osmium tetroxide. The precipitates retained their integrity and distribution through the preparation procedure and were successfully analyzed with a variety of electron beam analytical instruments. We have used conventional formaldehyde fixation, ethanol dehydration, and sections of paraffin-embedded rat and dog lung to localize insoluble europium particles (Leach et al., 1973). An example of such a preparation is seen in Fig. 8.9.

Analysis of lung tissue for asbestosis bodies has been accomplished us-

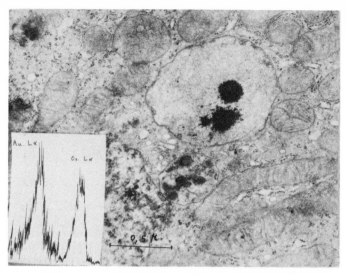

Fig. 8.8 Transmission electron micrograph of thin section of osmium-fixed kidney, from rat treated with gold salts. Insoluble particulate deposits are seen within a mitochondrion. The X-ray spectrum of the deposit is displayed in the inset. It can be seen that the particles are composed of gold and osmium. (Courtesy of P. Galle.)

ing preparation techniques common in pathological studies (von Rosenstiel and Zeedijk, 1968; Banfield *et al.*, 1969). Langer and associates have been very active in the development of methods to identify and analyze asbestos bodies in lung tissue. Some of these methods have been described and evaluated in recent reviews (Langer and coworkers, 1972a and b). They receive lung tissue from various sources and the tissue is usually fixed in a formaldehyde fixative, dehydrated, and embedded in paraffin according to the protocol in use at the institution providing the specimen. They reject samples that have been fixed in solutions containing heavy metals (e.g., $HgCl_2$, $KMnO_4$) which might interfere with analysis. These steps do not appear to permit the loss of any asbestos, presumably because of its great insolubility and the nature of its attachment to the lung tissue.

Three main types of preparation procedure are employed. In some cases, the paraffin-embedded lung specimens are sectioned and the sections are placed on glass slides for preliminary examination with the light microscope. A micromanipulator is used to remove individual asbestos fibers which are then placed on an epoxy film which is in turn supported on an electron microscope grid. The fibers are then coated with a thin layer of vacuum evaporated carbon and analyzed.

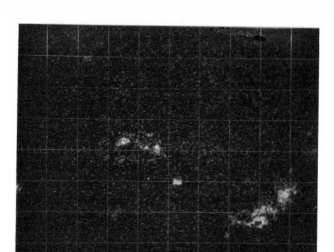

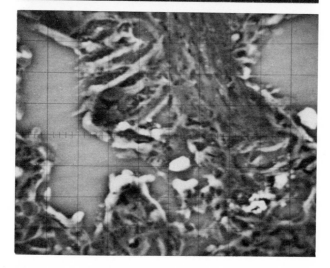

Fig. 8.9 Sample current and europium (Lα) X-ray images of 2 μm thick section of lung from dog with long-term exposure to europium aerosol. The tissue was fixed and sectioned according to conventional histological procedures. Substantial amounts of insoluble europium deposits are seen in the interstitial tissue of the lung. 10 μm/screen division.

In order to obtain fibers from larger amounts of tissue than a single section, they have developed methods for "bulk" analysis. One method involves cutting sections 25 μm thick, which are pooled by being stacked one on top of the other to a height of ~175 μm. The entire stack is then ashed at low temperature with nascent oxygen (Thomas, 1969), and the resulting particles are dispersed on a substrate prior to analysis. Other "bulk" samples have been ashed by exposure to 40% KOH at 100°C for 1 hr. The freed particles were collected by centrifugation, and washed three times by being suspended in distilled water, each time being collected by centrifugation. They are then dispersed on a suitable substrate and analyzed. In order to test these methods, these authors have treated asbestos fibers of standard composition according to the same methods and analyzed them. They found no alteration in composition produced by the method.

A method for the preparation of particle-containing tissue that permits examination of relict tissue and its incorporated inorganic materials in both electron probe and electron microscope has also been described (Langer and Pooley, 1972b). A 6μm thick paraffin section of lung mounted on a glass slide is placed in xylene for ~30 sec to remove the paraffin, into ethanol for ~30 sec to dilute the xylene, and then is permitted to dry in air. The authors recommend using the top of a Petri dish to contain the xylene and ethanol, and fresh xylene and ethanol for each section to prevent crosscontamination. The dried slide is placed on a ceramic plate and heated to 450°C in a muffle furnace. The time required for ashing may vary from 10 to 30 min; the preparation may be removed from the furnace when the section appears ash-gray in color.

The glass slide is allowed to cool, and the ashed remains of the section are outlined with a mask of cellophane tape. A 5–10% by weight solution of polyvinyl alcohol (PVA) is used to cover the area outlined by the mask, and must overlap the cellophane tape mask. The PVA–water mixture must be autoclaved at 130°C and several atmospheres of pressure to ensure complete solution. The PVA is allowed to dry overnight at room temperature. Forceps are used to pull the hardend PVA from the slide, using the cellophane tape as a peeling "vehicle." All the ashed residue pulls away from the slide and remains with the hardened PVA. The peeled plastic is inverted and taped again to the glass slide. This is coated with a heavy layer of carbon in a vacuum evaporator. The specimen is then scored with a scalpel to produce pieces sufficiently small to fit on an electron microscope grid.

Distilled water in a beaker is brought to a boil, then removed from the heat and allowed to stop agitating. The scored specimen is placed, PVA side down, on the surface of the water, and the beaker is covered with a Petri dish. When beaker and water have reached room temperature the PVA should have completely dissolved. The tape mask is pulled under the water with forceps, leaving the carbon film and attached residue of ash floating on the surface. The specimen will tend to separate along the score lines. Each piece may be picked up on an electron microscope grid. "Locator" type grids are recommended to facilitate locating the same area in electron microscope and electron probe. An edge of filter paper is used to remove excess water from specimen and forceps, care being taken not to damage the delicate specimen. Any remaining moisture may be removed by holding the grid several inches away from a Bunsen burner for several seconds.

The advantages of this technique are the production of uncoated asbestos fibers that can be rapidly identified in the electron microscope and then analyzed with the electron probe. Langer reports that this method has substantially speeded up their entire investigation. While this technique is very useful for material as insoluble as asbestos, it is unlikely that it can be used as described for more soluble materials.

Bone

Bone was one of the first biological samples examined by electron probe analysis. The relatively high density of bone, at least compared to soft biological materials, made it similar to the many metallurgical and mineral samples that had been successfully analyzed, and methods comparable to that used for hard, nonbiological samples could be readily adapted. Thus, thick blocks of bone that completely contain the electron beam could be analyzed. Mellors (1966) tried several different, but unspecified, methods of preparing bone for analysis before deciding on the one reported. The procedure finally adopted was to use normal human bone specimens that had been frozen and stored at −70°C.

The frozen bones were cut into blocks about 0.3–1.5 cm² in cross section and 0.03 cm thick, with a diamond or jeweler's saw. Fixation was in absolute alcohol for 4 hr, and the samples were "defatted" overnight in a 3 : 1 mixture of absolute alcohol and chloroform. Lucite powder was added to the chloroform phase of the specimens, and the chloroform was allowed to evaporate over a period of several days. The lucite-infiltrated specimens were then embedded in cylinders of lucite with a hot-stage hydraulic press. The face of the specimen was then polished with several grades of silicon carbide abrasive papers, absolute ethanol being used as lubricant. The final polish was produced with diamond paste on cloth. After polishing, the blocks were cut into smaller cubes $\frac{1}{2}$ cm on a side, and cleaned with ethanol in an ultrasonic cleaner. A 40 μm thick layer of carbon was vacuum-deposited on the samples to make them conductive.

This method was aimed at preserving the mineral portion of the bone that was not soluble in any of the solvents employed. Obviously any elements held in place by association with lipid components, or any elements soluble in ethanol or choroform, are likely to be lost during processing; this loss is in amounts proportional to solubility in and length of exposure to the solvents.

In this case the mineral phase of bone had been well characterized by many previous bulk analyses. Thus the mean of electron probe analyses of microscopic portions of bone could be referred to values obtained by bulk chemical analysis. The fact that this analysis resulted in mean calcium to phosphorus ratios within the range reported on bulk analyses suggested that the preparatory method was adequate for the purpose of studying these elements.

The preparation of a hard tissue such as bone is not without hazards. Boyde and Switsur (1963) have used scanning electron microscopy and interference light microscopy to elucidate the surface of polished bone samples and sections cut with diamond knives. They point out that ordinary polishing procedures may remove more material from softer regions than from harder ones, resulting in a rough surface for analysis. Such a rough surface may show differential absorption, with X-rays emitted from pits in soft regions being absorbed by peaks of surrounding hard material; X-rays emitted from peaks of hard material will not be absorbed proportionately.

There may be other difficulties resulting from the polishing procedure. Boyde and Switsur (1963) suggest that ordinary procedures may damage what in bulk analyses would ordinarily be considered a "surface" layer, but in electron probe analysis might constitute a layer as deep as that excited by the electron beam. They suggest that infiltrating the specimen with plastic may compensate for surface distortions. These authors emphasize that the junctions between hard and soft tissues are especially liable to distortion in ordinary polishing procedures. Unless the soft tissue is supported during polishing, e.g., by infiltration with methacrylate or other polymers, the soft tissues may be torn out or may collapse onto the hard tissue.

Brooks et al., (1962) used a preparatory method standard for microscopy to obtain sections of cartilage. Epiphyseal cartilage was fixed for 2 hr at room temperature in 10% formaldehyde buffered with veronal acetate to pH 7.35. Dehydration was with graded acetone–water mixtures, and the embedding material was 20/80 methyl/butylmethacrylate mixture. Sections 2 μm in thickness were cut with glass knives, and mounted on quartz slides. The embedding material was removed with xylene and the section was coated with carbon. The sample was analyzed only for calcium, and a calcium concentration differential was detected in calcified vs. uncalcified areas. The authors were careful to point out that their results only described the distribution of calcium bound tightly enough to resist removal during preparation.

To some extent the same limitations would apply to the preparations analyzed by Remagen et al., (1969). These authors employed 1% OsO_4 at pH 7.2 (with an unspecified buffer) at 4°C to fix metaphyseal portions of rat femurs for 2 hr. The sample was dehydrated in acetone and embedded in Araldite. Thin sections were cut on an ultramicrotome. In cutting some sections, they filled the boat with a saturated solution of tertiary calcium phosphate. Boothroyd (1964) had recommended this technique to prevent loss of calcium from the sections while they floated on the surface of the fluid in the boat. However, Remagen et al. (1969) found no difference between those sections cut onto distilled water and those cut onto the calcium phosphate solution. Nor did they find that staining with lead citrate or uranyl acetate influenced their results.

Although details of the preparation procedures were lacking because of journal space limitations, Carlisle (1970) reported the use of several preparative techniques for forming bones. She had reported the occurrence of silicon at calcifying sites and wished to avoid any possibility that the preparative technique could be responsible. Using the tibia from young mice and rats, samples were prepared by: (1) freeze-drying and embedding in polymer, (2) vacuum-drying and embedding in polymer, (3) hand-polishing of freeze-dried and vacuum-dried, polymer-embedded sections, (4) cryostat-cutting and subsequent freeze-drying, and (5) fixation with absolute ethanol and embedding in paraffin. Since the analyses were similar in each of preparations it may be concluded that each of these would produce specimens satisfactory for the analysis of mineralized and premineralized tissue.

Teeth have been prepared for analysis by methods similar to those used for bone. Embedding and polishing techniques are probably adequate for mineralized tissue, but soft tissues may be subject to the same hazards found in preparing bone (Frazier, 1966, 1967). In addition, Hall and Höhling (1969) have reported on a method to produce a freeze-dried section of tooth.

Soft Tissues

Ingram et al. (1972) have described a freeze-drying technique that was used successfully with several different soft tissues to preserve the intracellular distribution of electrolytes. Tissue is cut into blocks about 1 mm thick, or frozen in situ with liquid nitrogen. It is then quick-frozen in liquefied propane cooled by liquid nitrogen. Drying is carried out in a vacuum at a pressure of ~10–50 μm Hg maintained with a rotary vacuum pump for 2–4 weeks. During this time the temperature is maintained between $-60°C$ and $-85°C$. The tissue is then fixed with osmium vapor and embedded in Epon 826. Although all Epon contains a rather high chlorine concentration, the chlorine content of Epon 826 is among the lowest. Sections are cut about 3 μm thick with a dry steel knife and mounted on a quartz slide coated with a layer of vacuum-evaporated carbon about 20 nm thick. Ingram et al., (1972) point out that their procedure does not result in the formation of large ice crystals and that morphology is adequately preserved.

Robison and Davis (1969) used a Freon spray to freeze thyroid tissue, which was then frozen-sectioned at 10–14 μm in a cryostat-microtome. Sections were freeze-dried; fixed with 20% glutaraldehyde, 10% formalin, 10% acetic acid, and 60% methanol for an unspecified period of time and then freeze-dried; or warmed to room temperature and then fixed with the fixative mixture described above. The iodine distribution was similar in all methods of preparation. This may be due to the fact that iodine is bound to a rather large molecule which does not diffuse readily. Sodium distribution, on the other hand, was much more dependent on the preparation procedure and in some cases sodium was completely removed.

Lehrer (Lehrer, 1969, 1971; Lehrer, et al., 1970; Lehrer and Berkeley, 1972) has developed a method for the preparation of brain tissue. The tissue of interest is exposed surgically and a stream of Freon 12, chilled to $-150°C$ by liquid nitrogen, is applied. With the Freon stream still running, the frozen tissue is rapidly removed and placed in a cryostat. The tissue is sectioned at $-20°C$ with a steel knife and the sections are dried in a vacuum at $-40°C$.

Lehrer and coworkers took advantage of the large size (120 μm or more in diameter) of the cells they analyzed to corroborate the results of electron probe analysis. They microdissected some sections so that they could analyze cytoplasmic, nuclear, and extracellular fragments by flame photometry. The results of electron probe analysis and flame photometry agree quite well. The concentration of intracellular potassium is what was expected, i.e., 140 mM, but the concentration of intracellular sodium is much higher than was expected (Lehrer, 1972). Whether some of this sodium could have entered the cells during the preparative procedure remains to be determined (see, e.g., Farrant and Woolgar, 1972).

In any drying process, as well as in the specific case of freeze-drying as pointed out by Ingram et al. (1972), one must also be concerned regarding the changes that can occur as water is removed (Rebhun, 1972). Elements that are free in solution may be displaced as the drying process proceeds, producing a receding front in regions with relatively little structure, e.g., the lumen of kidney tubules. The elements left behind as water evaporates may be concentrated at the first physical barrier they encounter. This movement may produce the appearance of a concentration at certain membranes or surfaces. This possibility for movement is often illustrated by the example of drying a container of salt water. As water is removed, the salt does not remain suspended in space; it is concentrated at the receding solution surface, and eventually precipitates on the bottom and walls of the container.

Further examples of artifact that may occur with freezing and drying have been presented by Lehrer and Berkeley (1972). These authors froze gelatin solutions which were made up to contain various concentrations of different elements. The blocks were rapidly frozen, and sectioned while frozen in a cryostat. Some sections showed that the elements of interest had separated into a concentrated phase and were no longer homogeneously distributed throughout the gelatin. They term these "crystallization artifacts" and attribute their occurrence to slow freezing. It would seem that the drying process might also have some influence on their formation. Descriptions of several frozen-sectioning procedures that are variations of freeze-drying appear in a symposium recently reported in *Micron* (Lacy, 1971), and others are discussed by Echlin (1971).

Sakai *et al.* (1969) and Nikaido *et al.* (1972) have described a method they employed to detect silicon in certain regions of human brain. They start with formaldehyde-fixed brain obtained at autopsy. The composition of the formaldehyde fixative is not mentioned, but the material had been stored in the fixative for periods ranging from two months to nine years. Frozen sections were cut at unspecified thicknesses and stained with iodine or aldehyde fuchsin to identify *corporea amylacea*, distinct carbohydrate-rich bodies occurring in brain, and associated with certain pathologies. The sections were mounted on glass slides which were then attached to bakelite mounts. Alternatively, sections were mounted directly on aluminum foil to avoid interference from the silicon X-ray emitted by the glass slides. The sections were coated with 20 nm of carbon in a vacuum evaporator.

The sections were analyzed for calcium, phosphorous, sulfur, and silicon. It was thought that the phosphorous was present as a phosphoric ester in a polyglucosan, and thus would be retained through formaldehyde fixation and staining for carbohydrates (iodine and aldehyde fuchsin). The nature of the chemical bonds that permitted calcium, sulfur, and silicon to be retained through the preparative procedures is unknown.

In cases where only one or a few elements are of interest, precipitation techniques have been useful. Komnick (1962, 1969) proposed a method for the use of potassium pyroantimonate to localize sodium ions. The

188 PRINCIPLES AND TECHNIQUES OF ELECTRON MICROSCOPY

rationale of the method is based on the observation that pyroantimonate forms an insoluble precipitate with sodium. Bulger (1969) has investigated the use of this technique and found that pyroantimonate will also precipitate with calcium, magnesium, and barium. Klein *et al.*, (1972) report that the method is differentially sensitive, reacting with $10^{-6}M$ Ca^{2+}, 10^{-5} M Mg^{2+} and $\sim10^{-2}$ M Na^+. This eliminates quantitative comparisons among these three ions in tissue prepared this way. The use of this technique is further complicated by the observations of Bulger (1969) and Page (1969) that pyroantimonate does not appear to pass through intact cell membranes freely.

Aldehyde fixatives which produce cell membranes with relatively high dielectric constants (Carstensen *et al.*, 1969, 1971) do not seem to permit the pyroantimonate ion to enter cells, so that precipitation is mostly extracellular. Osmium tetroxide seems to be required to disrupt the integrity of the membrane and let the pyroantimonate ion enter the cell. Several slight variations of the technique have been employed, but the following seems typical: Lane and Martin (1969) used a mixture of 1% OsO_4 and 2% potassium pyroantimonate in 0.01 N potassium acetate buffer at a pH of 6.9. Fixation was for 1 hr followed by ethanol dehydration, infiltration with propylene oxide, and embedding in Epon 812.

Libanati and Tandler (1969), Tandler *et al.* (1970) and Kierszenbaum *et al.* (1971) disagree with the need for osmium disruption of plasma membranes and have used only potassium pyroantimonate as both fixative and precipitant. Their evidence leads them to conclude that the presence of osmium or an aldehyde fixative permits leakage of material from the cell, and subsequent precipitation with pyroantimonate in extracellular spaces. It seems likely that the question of whether pyroantimonate should or should not be used with osmium or an aldehyde fixative will only be answered with further work.

The procedure used by Tandler *et al.* (1970) begins with the preparation of a saturated aqueous solution of potassium pyroantimonate by boiling in deionized or double-glass-distilled water, and cooling rapidly to room temperature. This solution has an alkaline pH of ~9.2 and is centrifuged to remove precipitate. The resulting solution is used immediately or let stand overnight. Rat kidney was perfused with this solution through the renal artery at room temperature. The tissue turned white rapidly. The entire kidney was then dissected free, and cut into small pieces in a drop of the fixative solution. These pieces were transferred to a larger volume of fixative for ~3 hr at room temperature. The tissue was then hardened in a 5% solution of formaldehyde in the potassium pyroantimonate fixative for 6–24 hr at room temperature. Precipitate in this solution is removed by centrifugation prior to use.

Tissue may then be washed in distilled water and post osmicated, or may be treated with hot (90–95°C) potassium pyroantimonate for 5 min, rapidly cooled, and washed with ice-cold double-distilled water. The heat treatment is included to remove any potassium pyroantimonate. Other cation precipitates with pyroantimonate should not be dissolved. The tissue was dehydrated with graded cold ethanol solutions and embedded in Maraglas after treatment with propylene oxide. Sections were cut with glass knives, but whether wet or dry was not specified. Calcium, magnesium, and sodium have been detected at sites of pyroantimonate precipitation.

Clearly, the usefulness of this technique depends on the ability of the pyroantimonate ion to reach sites of cation localization before the cation can diffuse away from its normal location within the cell. The fact that the pyroantimonate is a large polar ion would suggest that it moves slowly through membranes and viscous cytoplasm. Tandler *et al.*, (1970) hypothesize that the sodium, calcium, and magnesium ions that precipitate with pyroantimonate are "loosely bound," probably to macromolecules or other fixed sites, and are not free to diffuse. When the pyroantimonate ion appears in the proximity of the ion, the affinity of the pyroantimonate ion for the cation is greater than that of the fixed site and the reaction with resultant precipitation occurs. If this mechanism is correct, the pyroantimonate may be a generally useful agent to preserve morphology as well as the distribution, though perhaps not the relative quantitation, of several intracellular ions.

Oxalate has also been used as a precipitating agent, mostly for calcium, since other plentiful physiological divalent ions do not form a complex as insoluble as calcium oxalate. Constantin *et al.* (1965) reported a procedure for perfusing muscle fibers with calcium followed by oxalate that produced dense deposits in terminal sacs of sarcoplasmic reticulum (Fig. 8.5).

Podolsky *et al.*, (1970) later identified these deposits as calcium oxalate using electron probe microanalysis. These investigators began by "skinning" the fibers, which involves removing the plasma membrane. This was necessary to permit oxalate to enter the cell, since the unfixed plasma membrane is normally impermeable to this ion. After perfusion with 10 mM $CaCl_2$ plus 80 mM sodium citrate, 10 mM sodium oxalate plus 140 mM KCl was added. Five minutes after the perfusion the preparation was fixed with 6.5% glutaraldehyde in 0.2 M cacodylate buffer (*p*H 7.2), washed in buffer, and postfixed in 2% osmium tetroxide. All of these solutions contained 10 mM oxalate. The material was rinsed in 10 mM oxalate, dehydrated in alcohol and acetone, and embedded in Araldite. Thin sections were cut conventionally. Both stained and unstained sections were examined, since lead staining was found to remove the oxalate deposits.

A similar method was used to study the distribution of calcium in two tissues, the chick chorioallantoic membrane and rat and chick small intestine, that actively transport this element (Coleman and Terepka, 1972b; Warner and Coleman, 1972b; Warner, 1972). In this case perfusion was not found to be necessary; instead, 6% acrolein (Carstensen et al., 1971) containing 1% potassium or sodium oxalate and buffered to pH 7.2 with 0.1 M cacodylate–HCl buffer, was used as the primary fixative and precipitant. Duration of fixation was 30 min at room temperature. Tissues were then rinsed in buffer, treated with 2% OsO_4 in the same oxalate-containing buffer for 30 min at room temperature, and dehydrated with ethanol which contained oxalate. Tissues to be embedded in plastic were infiltrated with propylene oxide which contained oxalate, and then embedded in Araldite. Tissues embedded in paraffin were infiltrated with xylene followed by paraffin, and finally embedded in Paraplast (m.p. 59°C).

Sections were cut with dry steel and glass knives and mounted on either quartz disks coated with vacuum-evaporated aluminum, silicon wafers, or vitreous carbon disks. Araldite sections were attached to substrates by floating them on a drop of ethanol and evaporating the ethanol with mild heat. Paraplast sections were "expanded" by floating on the surface of a 5% oxalate solution at 55°C for a short (1–15 sec) period; they were then mounted on silicon wafers, and sections and mountings were then dried in air at room temperature. Paraplast was removed by immersion in xylene for 5–15 min. Figure 8.10 shows tissue prepared in this way. This method was shown to preserve tissue morphology even at the level of electron microscopic ultrastructure and to preserve ~85% of the total calcium. It produces no selective loss of calcium from the pool of calcium in transport or the calcium contained in the rest of the tissue. There is no evidence of calcium redistribution, nor of the formation of artifactual aggregates of crystalline or amorphous precipitate.

In order to study protein and sulfur accumulation in keratinizing skin, Sims and Hall (1968) fixed plantar skin from albino rats in 10% formaldehyde for 24 hr. These were embedded in paraffin "in the usual way." Sections 5 μm thick, perpendicular to the skin surface, were cut and placed on thin nylon film supports. Both surfaces of the sections were coated with an 80 nm thick layer of aluminum for analysis. Since the sulfur is bound by covalent bonds to macromolecules, they would be expected to be retained by this method.

The method of preparing the nylon film and mounting sections is of interest. A 2 ml drop of 10% nylon in isobutyl alcohol at 60°C is squirted onto a clean surface of distilled water. The film is removed from the sur-

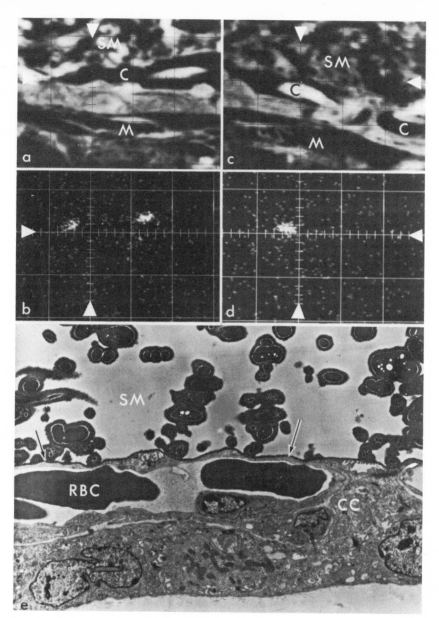

Fig. 8.10 Sample current and calcium X-ray images of thick sections from two chick chorioallantoic membrane preparations fixed in the presence of oxalate to preserve the distribution of calcium (a and b; c and d). Calcium localizations are found between the noncellular shell membrane (SM) and lumen of respiratory capillaries (c). Identical grid marks are marked by a triangle in each pair. M, mesoderm, 5 μm/screen division. A transmission electron micrograph ($\times$5,800) of a thin section from a membrane prepared the same way appears (e) to show that the normal structural relationships are not disrupted by this procedure. CC is a "capillary-covering" cell which extends long cytoplasmic processes (arrow) between the shell membrane and respiratory capillaries. RBC, erythrocyte. (From Coleman and Terepka, 1972b.)

face of the water with a wire frame and dried in a dessicator. One end of a 1 cm diameter aluminum tube is covered with the film to make a carrier for sections. The ribbon of paraffin-embedded sections is flattened on a water bath, cooled, and dried on filter paper. The paraffin is removed with xylol and the sections brought into water through descending grades of alcohol. They are transferred in water to the nylon carriers and dried in a dessicator. The use of the nylon substrate provides the advantage of an electron-transmitting substrate. This is required for quantitation by the Hall method (1971), and may also improve spatial resolution by reducing electron backscatter from an opaque substrate (cf. Hall *et al.*, 1972).

Goldfischer and Moskal (1966) used the electron probe to demonstrate that copper accumulated in liposomes in Wilson's Disease. The technique consisted of fixing biopsy samples in 3% glutaraldehyde for 2 hr in cacodylate buffer (pH 7.4). The tissue was then washed overnight in buffer and frozen sections (7.5 μm thick) were cut with a freezing-stage microtome. The sections were incubated in a Gomori lead medium with β-glycerophosphate or cytidine-5′-monophosphate as substrate. Control sections had substrate omitted from the incubation mixture. After incubation the sections were air-dried on glass slides, coated with carbon, and analyzed for lead and copper. The success of this technique depended on the formation of insoluble lead phosphate as a result of enzyme activity as well as on the copper in the tissue being present in a form so insoluble not to be removed during the processing.

Single Cells and Isolated Orangelles

Andersen (1967) reported the use of simple air drying as a method to preserve the contents of *Amphiuma* blood cells. The sodium and chlorine X-ray intensities from these cells was significantly greater than those from cells which were freeze-dried or from cells which were fixed in glutaraldehyde and osmium, embedded in plastic, and sectioned. However, it is not possible from his data to decide whether this sodium corresponded to the amount that occurred *in vivo* or whether it might have precipitated onto the cell surface or entered the cell during drying.

Carroll and Tullis (1968) used a simple smear method to dry blood cells on polished silicon disks, a technique which amounts to air-drying a thin layer of cells. They examined several populations of cells and reported that leucocytes of tumor origin often contained abnormally high concentrations of titanium and zinc. This finding was, however, disputed by Beaman *et al.* (1969), who claimed that titanium and zinc were restricted to noncellular debris. These authors pointed out that the drying process distorted the characteristic morphological features which are commonly used to identify blood cells. Furthermore, the cells had been dried on an opaque substrate, which meant that they had to be identified by means of reflecting optics. The images produced in reflecting microscopes are not familiar to most hematologists. Finally, Beaman *et al.* (1969) point out that simple air-drying of cells subjects them to the hazards of airborne contamination; they show the importance of comparing electron-induced (e.g., sample current, backscattered electron, or secondary electron) images with X-ray images (see also Coleman and Terepka (1972a)).

Recently, Kimsey and Burns (1973) have critically examined the preparation of blood smears by air-drying. They found that the medium in which the cells are suspended while drying can influence the results. For example, cells dried in $CaCl_2$ solutions appeared to contain substantial amounts of calcium. Apparently, the presence of calcium is due to the adsorption of ions onto the surfaces of the cells, for rapid washing removes it. Suspending the cells in isotonic sucrose before drying can eliminate the interference by extracellular ions, so that intracellular constituents can be validly analyzed.

Coleman *et al.*, (1972; 1973a and b) have prepared protozoan cells using a similar technique. This method was devised to analyze lipid granules that occur within the cytoplasm of these cells. The process of heat-fixation retains recognizable morphology and the normal distribution of diffusible elements. The cells are placed on a silicon disk in a drop of culture medium. After a minute, during which time the cells settle to the surface of the disk, most of the medium is removed, either with a Pasteur pipette or the torn edge of a piece of filtered paper. The silicon disk is immediately moved at moderate speed through the flame of a propane torch, to dry the cells rapidly.

Subsequent analysis shows that potassium is restricted to the cell body and has not leaked out during preparation. This is taken as an indication that the cell membrane remained intact during the drying and the intracellular contents have remained within the cell. In the two protozoa examined in these papers, the cytoplasm contained lipid droplets rich in cations. This permitted a test of whether redistribution might have occurred within the cytoplasm. By comparing the sample current images and X-ray images of the soluble cations of the droplets it could be seen that the two were identical. This indicated that no detectable diffusion of cations had occurred out of the droplets into the surrounding cytoplasm.

Diffusion from the cytoplasm into the granules was ruled out by examining the sample current and X-ray profiles of the droplets and surrounding cytoplasm. Had cations diffused into the droplets there would have been a drop in the concentration of these elements in the cytoplasm surrounding the droplets, and this decrease would have appeared in the X-ray profiles. X-ray profiles showed no evidence of such a "halo" of decreased concentration surrounding the sample current image profile. Thus, there seemed to be no evidence that the rapid drying caused substantial redistribution of soluble, diffusible ions. This same procedure as well as air-drying have been used to prepare chick red blood cells for analysis of their sodium and potassium content, and to determine whether sodium had leaked into cells isolated from chick intestinal epithelium (Barrett and Coleman, 1973). Cells prepared by these methods are seen in Figs. 8.11 and 8.12.

This technique is also suitable for isolated cell organelles. By using electron paramagnetic resonance, the ability of mitochondria to incorporate manganese has been studied (Gunter and Puskin, 1972; Puskin and Gunter, 1972). When isolated mitochondria are incubated with manganese under various conditions, they accumulate different amounts of manganese. When these mitochondria are prepared by heat fixation, they can be analyzed individually with the electron probe. Figure 8.13 shows the sample current image of such a preparation. Table 8.2 presents the results

Table 8.2 Electron Probe Analyses of Isolated Rat Liver Mitochondria*

	Low Mn ($N = 18$)	*High* Mn ($N = 20$)
Mn	0.472 ± 0.145	4.365 ± 0.856
P	0.558 ± 0.138	3.964 ± 0.805
Mn/P	0.733 ± 0.025	1.119 ± 0.162

* According to the BICEP procedure (Warner, 1972). The mitochondria were incubated with manganese either in the absence of a permeant ion (low Mn), or the presence of a permeant ion (high Mn). The results are presented as atomic percentages, analogous to mole ratios. The manganese concentrations in atomic percentage correspond to a molar concentration of 273 μmole/g protein (low Mn) and 2500 μmole/g protein, values similar to those estimated by electron paramagnetic resonance methods (Gunter and Puskin, 1972; Puskin and Gunter, 1972).

of the analysis of two different preparations, one which accumulated only small amounts of manganese, and one which accumulated many times as much. The mean amounts of manganese accumulated by the mitochondria correspond to that found by electron paramagnetic resonance: greater than 70 nmoles/mg protein in the low Mn preparation and about 2,600 nmoles/mg protein in the high Mn preparation (Gunter and Puskin, 1972; Puskin and Gunter, 1972).

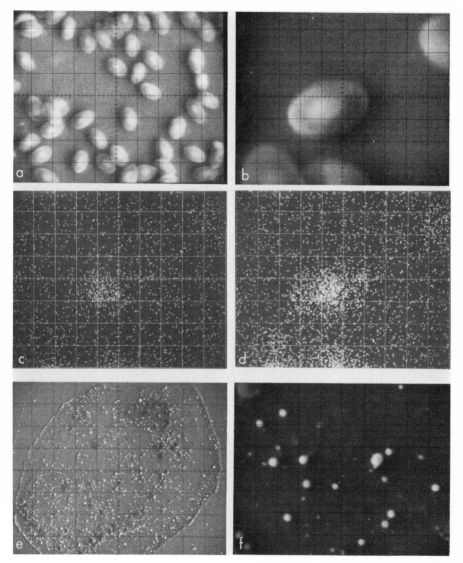

Fig. 8.11 Secondary electron and X-ray images of air-dried erythrocytes. (a) 10 μm/screen division. (b) 2 μm/screen division. (c) Phosphorous $K\alpha$ X-ray image of cells in (b) showing this element most prominent in the region of the nucleus. (d) Potassium X-ray image, showing this element distributed throughout the cell. Parts (c) and (d) were 15-min exposures, so that the random background radiation from the silicon disk substrate (2 cps) is rather prominent (courtesy of E. Barrett). Parts (e) and (f) are secondary electron images of a heat-fixed *Amoeba proteus*. The preparation appears as a collapsed balloon containing spherical "refractive bodies." (e) 100 μm/screen division. (f) 8 μm/screen division.

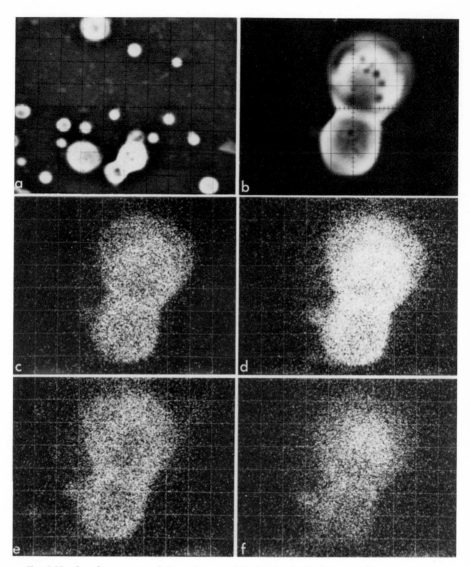

Fig. 8.12 Sample current and X-ray images of "refractive bodies" in *Amoeba proteus*. (a) 6 μm/screen division. (b) 1 μm/screen division. (c) Calcium Kα X-ray image of refractive body in b. (d) Phosphorous Kα image, (e) potassium Kα image, and (f) magnesium Kα image of the same refractive granule (Coleman *et al.* 1973a).

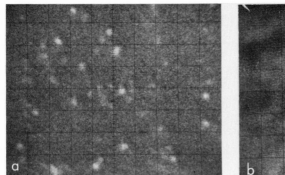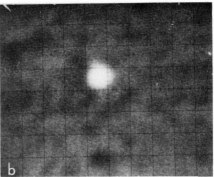

Fig. 8.13 Sample current images of heat-fixed mitochondria isolated from rat liver and incubated with manganese (Gunter and Puskin, 1972; Puskin and Gunter, 1972).

The success of this preparative method may be due to the fact that most cellular membranes are relatively permeable to water and the drying may occur as a type of "flash evaporation," in which water is rapidly sublimed through the membranes while other elements with lower vapor pressures, to which the membranes are less permeable, remain behind.

Plant Tissues

Plant tissues have been examined by several workers. Läuchli and Lüttge (1968) embedded plant tissue in brain tissue from experimental animals, and used cryostat sections (~8 μm thick) of frozen plant tissue (*Pisum sativum*) mounted on silicon disks. Kaufman *et al.* (1969) used hand-cut slices (2–4 layers thick) of *Avena* which were freeze-dried in a commercial freeze-drying apparatus. The dried slices were then analyzed.

Rasmussen (1969) prepared *Zea mays* by two methods, paraffin embedding and cryostat sectioning. Tissues for paraffin embedding were fixed in formalin–acetic-acid–alcohol, but the concentration of each component was not given. They were then dehydrated and embedded in paraffin, and 12 μm thick sections were cut and mounted on carbon disks with Haupt's adhesive. Prior to analysis the paraffin was removed with xylene. Tissues for cryostat cutting were frozen and mounted in "Optimum Cutting Compound" (O.C.T. $-15°C$ to $-30°C$, Fisher Scientific Company). Sections 16 μm thick were cut at $-16°C$, placed on polished carbon disks, and allowed to dry at room temperature. Rasmussen observes that although the cryostat method retains elements in their exact location, it does not preserve cellular detail as well as the paraffin embedding method.

Subsequently, Waisel *et al.* (1970) used a somewhat different method to prepare *Phaseolus vulgaris* and *Hordeum sativum*. Root tip was embedded either in Lipshaw's M-1 (Lipshaw Scientific Company) embedding matrix or in fresh brain of experimental animals according to the technique of Läuchli (1967a and b) and then frozen in cold acetone at −70°C. Sections 20 μm in thickness were cut with a cryostat at −15°C and placed on cold nickel plates. The nickel plates were kept cold and placed in a lyophilizer until the sections were dry, resulting in sections that had been kept frozen until dry. These authors concluded that the technique of keeping the tissue frozen until dry had resulted in preservation of the distribution of labile elements superior to that reported by Rasmussen (1969).

Läuchli *et al.* (1970) have reported a method for freeze-substitution with anhydrous ethylether followed by embedding in Spurr's (1969) low-viscosity medium. Sections are cut onto a trough filled with hexylene glycol, which is effective in preventing loss of water-soluble materials from the sections.

Microincineration

Thomas (1969) has shown that microincineration, either at high temperatures or at low temperatures with nascent oxygen, can be a useful ancillary technique for microanalysis (Fig. 8.14). He has used these techniques to remove organic materials from preparations made by drying suspensions of isolated cell organelles, and from relatively thick sections of material embedded in plastic. The major advantage of the technique is that by removing background, it increases the concentration of nonvolatile elements so that they may be detected more easily, while the ash pattern permits identification of many cell structures. The use of low-temperature incineration has been shown to result in less movement of elements during the ashing, and a greater retention of nonvolatile elements. For a detailed discussion on ultramicroincineration, the reader is referred to Hohman's chapter in this volume.

This paper is based in part on work performed under contract with the US Atomic Energy Commission at the University of Rochester Atomic Energy Project and in part on work supported by USPHS Research Grant AM-14272, and has been assigned Univ. of Rochester Atomic Energy Project Report No. UR-3490-288.

Review of the literature was greatly simplified by the diligent labor of

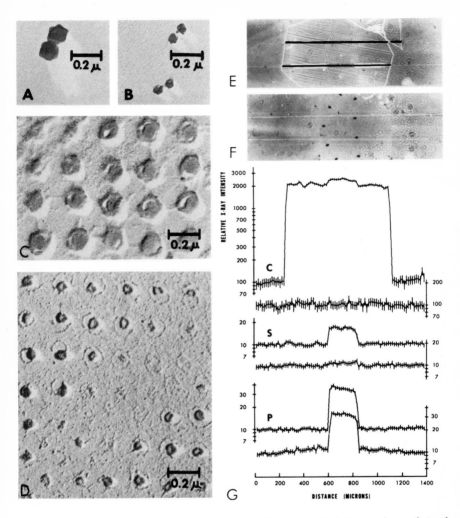

Fig. 8.14 Low-temperature (oxygen plasma) microincineration and electron probe analysis of *Tipula* iridiscent virus (TIV). (A and B) Transmission electron micrographs of shadowed preparations of TIV before and after incineration, respectively. The ash is largely due to the phosphate residue of the viral DNA. (C and D) Transmission electron micrographs of embedded, shadowed preparations of virus crystals before and after ashing, respectively. (The embedding material was removed prior to shadowing to improve contrast.) (E and F) A 1 μm thick section of methacrylate-embedded TIV crystal before and after ashing, respectively. The horizontal lines indicate the path of the electron beam in recording the X-ray profiles in (G). The small arrows in (F) indicate the outlines of the remains of the virus crystal after ashing. (Courtesy R. S. Thomas.)

T. A. Hall, Cambridge, who has organized an up-to-date bibliography of probe literature and made it available to electron probe users.

References

Andersen, C. A. (1967). An introduction to the electron probe X-ray micro-analyzer and its application to biochemistry. *In:* Methods of Biochemical Analysis XV (Glick, D., Ed.), pp. 147–270. Interscience Publishers, New York.

———, and Hasler, M. F. (1966). Extension of electron microprobe tech-niques to biochemistry by the use of long wavelength X-rays. *In:* X-Ray Optics and Microanalysis (Castaing, Descamps and Philibert, Eds.), pp. 310–327. Hermann, Paris.

Anderson, T. F. (1966). Electron microscopy of microorganisms. *In:* Physical Techniques in Biological Research, Vol. 3 (Pollister, A. W., Ed.), pp. 319–387. Academic Press, New York.

Appleton, T. C. (1971). Dry ultrathin frozen sections for electron microscopy: The cryostat approach. *Micron,* **3,** 101.

Bacaner, M., Broadhurst, J., Hutchenson, T., and Lilley, J. (1972). High resolution localization of ions correlated with electron optical image of muscle sarcomeres by analysis of non-dispersive X-rays generated by elec-tron bombardment in scanning electron microscope. (Abstr.) *Fed. Proc.,* **31,** 324.

Bahr, G. F., Johnson, F. B., and Zeitler, E. (1965). The elementary compo-sition of organic objects after electron irradiation. *Lab. Invest.,* **14,** 377.

Banfield, W. G., Tousimis, A. J., Hagerty, J. C., and Padden, T. R. (1969). Electron probe analysis of human lung tissues. *In:* Progress in Analytical Chemistry, Vol. III (Earle, K. M., and Tousimis, A. J., Eds.), pp. 6–34. Plenum Press, New York.

Banfield, W. G., Grimley, P. M., Hammond, W. G., Taylor, C. M., de Florio, B., and Tousimis, A. J. (1971). Electron probe analysis for iodine in human thyroid and parathyroid glands, normal and neoplastic. *J. Nat. Cancer Inst.,* **46,** 269.

Bank, H., and Mazur, P. (1973). Visualization of freezing damage. *J. Cell Biol.,* **57,** 729.

Beaman, D. R., and Isasi, J. A. (1970). A critical examination of computer programs used in quantitative electron probe analysis. *Anal. Chem.,* **42,** 1540.

Beaman, D. R., Nishayama, R. H., Penner, J. A. (1969). The analysis of blood diseases with the electron microprobe. *Blood,* **34,** 401.

Barrett, E. J., and Coleman, J. R. (1973). Sodium and potassium content of single cells: Effects of metabolic and structural changes. *Proc. 8th Nat. Conf. Electron Probe Analy.,* New Orleans.

Bernhard, W., and Leduc, E. H. (1967). Ultrathin frozen sections. I. Methods and ultrastructural preservation. *J. Cell Biol.,* **34,** 757.

Birks, L. S. (1971). Electron Probe Microanalysis, 190 pp. Wiley–Interscience, New York.

Boothroyd, B. (1964). The problem of demineralization in thin sections of fully calcified bone. *J. Cell Biol.*, **20**, 165.

Boyde, A., and Switsur, V. R. (1963). Problems associated with the preparation of biological specimens for microanalysis. *In:* X-Ray Optics and X-Ray Microanalysis (Proc. 3rd intern. symp. X-ray optics X-ray microanal.) (Pattee, Jr., H. H., Coslett, V. E., and Engström, A., Eds.), pp. 499–506. Academic Press, New York.

Brooks, E. J., Tousimis, A. J., and Birks, L. S. (1962). The distribution of calcium in the epiphyseal cartilage of the rat tibia measured with the electron probe X-ray microanalyzer. *J. Ultrastruct. Res.*, **7**, 56.

Bulger, R. E. (1969). Use of potassium pyroantimonate in the localization of sodium ions in rat kidney tissue. *J. Cell Biol.*, **40**, 79.

Carlisle, E. M. (1970). Silicon: A possible factor in bone calcification. *Science*, **167**, 279.

Carroll, K. G., and Tullis, J. L. (1968). Observations on the presence of titanium and zinc in human leucocytes. *Nature*, **217**, 1172.

Carstensen, E. L., Coopersmith, A., Ingram, M., and Child, S. Z. (1969). Stability of erythrocytes fixed in osmium tetroxide solutions. *J. Cell Biol.*, **42**, 565.

Carstensen, E. L., Aldridge, W. G., Child, S. Z., and Sullivan, P. (1971). Stability of cells fixed with glutaraldehyde and acrolein. *J. Cell Biol.*, **50**, 529.

Castaing, R. (1951). Application des sondes electroniques à une methode d'analyse ponctuelle chimique et cristallographique. Dissertation, Univ. of Paris. ONERA Publication No. 55.

Christensen, A. K. (1971). Frozen thin sections of fresh tissue for electron microscopy, with a description of pancreas and liver. *J. Cell Biol.*, **51**, 772.

Cohen, A. L. (1974). Critical point drying. *In: Principles and Techniques of Scanning Electron Microscopy: Biological Applications*, Vol. 1 (Hayat, M. A., Ed.). Van Nostrand Reinhold Company, New York and London.

Colby, J. W. (1968). Quantitative microprobe analysis of thin insulating films. *In:* Advances in X-ray Analysis, Vol. 11 (Nekirk, J. B., Mallett, G. R., and Pfeiffer, H. G., Eds.), pp. 287–305. Plenum Press, New York.

Coleman, J. R., Nilsson, J. R., Warner, R. R., and Batt, P. (1972). Qualitative and quantitative electron probe analysis of cytoplasmic granules in *Tetrahymena pyriformis. Exptl. Cell. Res.*, **74**, 207.

——— (1973a). Electron probe analysis of refractive bodies in *Amoeba proteus. Exptl. Cell Res.*, **76**, 31.

——— (1973b). Effects of calcium and strontium on divalent ion content of refractive granules in *Tetrahymena pyriformis. Exptl. Cell Res.*, **80**, 1.

Coleman, J. R., and Terepka, A. R. (1972a). Electron probe analysis of the calcium distribution in cells of the embryonic chick chorioallantoic membrane. I. A critical evaluation of techniques. *J. Histochem. Cytochem.*, **20**, 401.

Coleman, J. R., and Terepka, A. R. (1972b). Electron probe analysis of the calcium distribution in cells of the embryonic chick chorioallantoic membrane. II. Demonstration of intracellular location during active transcellular transport. *J. Histochem. Cytochem.*, **20**, 414.

Constantin, L. L., Franzini-Armstrong, C., and Podolsky, R. J. (1965). Localization of calcium accumulating structures in striated muscle fibers. *Science*, **147**, 158.

Cosslett, V. E., and Duncumb, P. (1956). Microanalysis by a flying-spot X-ray method (letter). *Nature*, **177**, 1172.

Echlin, P. (1971). The examination of biological materials at low temperatures. *In:* Scanning Electron Microscopy/1971. Part I. (Proc. 4th Ann. Scanning Electron Microsc. Symp.), pp. 225–232. IIT Research Institute, Chicago, Ill.

Farrant, J., and Woolgar, A. E. (1972). Human red cells under hypertonic conditions: A model system for investigating freezing damage. I. Sodium chloride. *Cryobiology*, **9**, 9.

Frazier, P. D. (1966). Electron probe analysis of human teeth: Some problems in sample preparation. *Norelco Reporter*, **13**, 25.

——— (1967). Electron probe analysis of human teeth (Ca/P ratios in incipient carious lesions). *Arch. Oral Biol.*, **12**, 25.

Galle, P. (1964a). Etude comparée au microscope électronique et à la microsonde du rein pathologique. *J. Microscopie*, **3**, 355.

——— (1964b). Oxalose rénal expérimentale. Etude au microscope électronique et par spectrographie des rayons X. *Nephron*, **1**, 158.

——— (1967a). Les nephrocalcinoses: Nouvelles données d'ultrastructure et de microanalyse. *Actual. Nephrol. Hop. Necker.*, 303–315.

——— (1967b). Microanalyse des inclusions minerales du rein. *Proc. 3rd. Intern. Cong. Nephrol., Washington, D.C.*, **2**, 306.

———, and Morel-Maroger, L. (1965). Les lésions rénales du saturnisme humain et expérimental. *Nephron*, **2**, 273.

———, Vivier, E., and Petitprez, A. (1968). Etude par microscopie électronique couplée à la spectrographie des rayons X d'inclusions de matière inerte chez *Spirostomum ambiguum. In:* Electron Microscopy 1968 (Proc. 4th Eur. Regional Conf.), **2**, 439–440. Tipografia Poliglotta Vaticana, Rome.

Goldfischer, S., and Moskal, J. (1966). Electron probe microanalysis of liver in Wilson's disease. *Amer. J. Pathol.*, **48**, 305.

Gunter, T. E., and Puskin, J. S. (1972). Manganous ion as a spin label in studies of mitochondrial uptake of manganese. *Biophy. J.*, **12**, 625.

Hall, T. A. (1971). The microprobe assay of chemical elements. *In:* Physical Techniques in Biological Research, Vol. 1 (Oster, G., Ed.), Part A, 2nd ed., pp. 157–275. Academic Press, New York.

———, and Höhling, H. J. (1969). The application of microprobe analysis to biology. *In:* X-ray Optics and Microanalysis (Mollenstedt, G., and Gaukler, K. H., Eds.), pp. 582–591. Springer-Verlag, Heidelberg.

——— Röckert, H. O. E., and Saunders, R. L. deC. H. (1972). X-ray Microscopy in Clinical and Experimental Medicine, 320 pp. Charles E. Thomas, Springfield, Ill.

Hayat, M. A. (1970). Principles and Techniques of Electron Microscopy: Biological Applications. Vol. 1, 412 pp. Van Nostrand Reinhold, New York.

———, and Zirkin, B. R. (1973). Critical point drying. *In:* Principles and Techniques of Electron Microscopy: Biological Applications, Vol. 3 (Hayat, M. A., Ed.). Van Nostrand Reinhold Company, New York and London.

Heinrich, K. F. J. (1968). Quantitative electron probe analysis. NBS Special Publ. 298, 299 pp.

Hodson, S., and Marshall, J. (1971a). Ultracryotomy: A technique for cutting ultrathin sections of unfixed frozen biological tissues for electron microscopy. *J. Microscopy,* **91,** 105.

——— (1971b). Migration of potassium out of electron microscope specimens. *J. Microscopy,* **93,** 49.

Ingram, M. J., and Hogben, C. A. M. (1968). Procedures for the study of biological soft tissue with the electron microprobe. *In:* Developments in Applied Spectroscopy, Vol. 6 (Baer, W. K., *et al.,* Eds.); p. 43. Plenum Press, New York.

Ingram, F. D., Ingram, M. J., and Hogben, C. A. M. (1972). Quantitative electron probe analysis of soft biologic tissue for electrolytes. *J. Histochem. Cytochem.* **20,** 716.

Kaufman, B. P., Bigelow, W. C., Petering, L. B., and Drogos, Z. B. (1969). Silica in developing epidermal cells of *Avena* internodes: Electron probe analysis. *Science,* **166,** 1015.

Kierszenbaum, A. L., Libanati, C., and Tandler, C. (1971). The distribution of inorganic cations in mouse testis. Electron microscope and microprobe analysis. *J. Cell Biol.,* **48,** 314.

Kimsey, S. L., and Burns, L. C. (1973). Electron probe microanalysis of cellular potassium distribution. *Ann. N.Y. Acad. Sci.,* **204,** 486.

Klein, R. L., Yen, S.-S., and Thureson-Klein, A. (1972). Critique on the K-pyroantimonate methods for semiquantitative estimation of cations in conjunction with electron microscopy. *J. Histochem. Cytochem.,* **20,** 65.

Komnick, H. (1962). Elektronenmikroskopische Lokalisation von Na+ und Cl− in Zellen und Geweben. *Protoplasma,* **55,** 414.

——— (1969). Histochemische Calcium-Lokalisation in der Skelettmuskulatur des Frosches. *Histochemie,* **18,** 24.

Lacy, D. (Ed.) (1971). First symposium on biological applications of combined high resolution electron microscopy and X-ray microanalysis. *Micron,* **3,** 81.

Lane, B. P., and Martin, E. (1969). Electron probe analysis of cationic species in pyroantimonate precipitates in Epon embedded tissue. *J. Histochem. Cytochem.,* **17,** 102.

Langer, A. M., Rubin, I. B., and Selikoff, I. J. (1972). Chemical characterization of asbestos body cores by electron microprobe analysis. *J. Histochem. Cytochem.,* **20,** 723.

————, and Pooley, F. D. (1972b). Chemical characterization of uncoated asbestos fibers from the lungs of asbestos workers by electron microprobe analysis. *J. Histochem. Cytochem.*, **20**, 735.

Läuchli, A. (1967a). Nachweis von Calcium-Strontiumablagerungen in Fruchsteil von *Pisum sativum* mit der Röntgen-Mikrosonde. *Planta,* **73**, 221.

———— (1967b). Untersuchungen über Verteilung und Transport von Ionen in Pflanzengeweben mit der Röntgen-Mikronsonde. I. Versuche an vegitativen Organen von *Zea mays. Planta,* **75**, 185.

————, and Lüttge, U. (1968). Untersuchung der Kinetik der Ionen-Aufnahme in des Cytoplasma von *Mnium*-Blattzellen mit Hilfe der Mikroautoradiographie und der Röntgen-Mikrosonde. *Planta,* **83**, 80.

————, Spurr, A. R., and Wittkopp, R. W. (1970). Electron probe analysis of freeze-substituted epoxy resin embedded tissue for ion transport studies in plants. *Planta,* **95**, 341.

Leach, L. J., Yuile, C. L., Hodge, H. C., Sylvester, G. E., and Wilson, H. B. (1973). A five year inhalation study with natural uranium dioxide dust. II. Postexposure retention and biologic effects in the monkey, dog and rat. *Health Phys.,* in press.

Lechène, C., Morel, F., Guinnebault, M., and de Rouffignac, C. (1969). Etude par microponction de l'élaboration de l'urine. I. Chez le rat dans differents états de diurèse. *Nephron,* **6**, 457.

Lehrer, G. M. (1969). The central nervous system sucrose space—a comparison of quantitative and auroradiagraphic data. *In:* Autoradiography of Diffusible Substances (Roth, L., ed.), p. 191. Academic Press, New York.

Lehrer, G. M. (1971). The determination of elemental composition in subcellular portions of individual neurons. *In:* Recent Advances in Quantitative Histo- and Cytochemistry (Dubach, V. C. and Schmidt, V. S., eds.), p. 183. Hans Huber, Bern.

Lehrer, G. M. (1972). Distribution of sodium and potassium in the nucleus and cytoplasm of neurons. *In:* Metabolic Compartmentalization in the Brain (Balázs, R., and Cremer, J. eds.), p. 259. The Macmillan Co., New York.

Lehrer, G. M., Katzman, R., and Wilson, C. (1970). Volume and density measurements in subcellular portions of single nerve cell bodies. *J. Histochem. Cytochem.,* **18**, 44.

Lehrer, G. M., and Berkeley, C. (1972). Standards for electron probe microanalysis of biologic specimens. *J. Histochem. Cytochem,* **20**, 710.

Libanati, C. M., and Tandler, C. J. (1969). The distribution of the water-soluble inorganic orthophosphate ions within the cell: Accumulation in the nucleus. Electron probe microanalysis. *J. Cell Biol.,* **42**, 754.

Mahler, H. R., and Cordes, E. H. (1971). Biological Chemistry, 2nd ed., 1009 pp. Harper and Row, New York.

Mellors, R. C. (1966). Electron microprobe analysis of human trabecular bone. *Clin. Orthoped. Related Res.,* **45**, 157.

Miller, E. L., and Wittry, D. B. (1966). Temperature controlled stage for electron probe microanalyzers. *Rev. Sci. Inst.,* **37**, 115.

Morel, F., de Rouffignac, C., Marsh, D., Guinnebault, M., and Lechène, C. (1969). Etude par microponction de l'élaboration de l'urine. II. Chez le *Psammomys* non diurétique. *Nephron*, 6, 553.

Neuman, W. F., and Neuman, M. W. (1958). The Chemical Dynamics of Bone Mineral, 209 pp. Univ. of Chicago Press, Chicago.

Nikaido, T., Austin, J., Trueb, L., and Rhinehart, R. (1972). Studies in aging of the brain. II. Microchemical analyses of the nervous system of Alzheimer patients. *Arch. Neurol.*, 27, 549.

Page, S. G. (1969). Structure and some contractile properties of fast and slow muscles of chicken. *J. Physiol.*, 205, 137.

Pearse, A. G. E. (1961). Histochemistry: Theoretical and Applied, 2nd ed., 998 pp. Little, Brown and Company, Boston.

Podolsky, R. J., Hall, T. A., and Hatchett, S. L. (1970). Identification of oxalate precipitates in striated muscle fibers. *J. Cell Biol.*, 44, 699.

Puskin, J. S., and Gunter, T. E. (1972). Paramagnetic resonance studies of a fraction of manganese showing hyperfine sextet within mitochondria. *Biophys. J.*, 12, 625.

Rasmussen, H. P. (1969). Entry and distribution of aluminum in *Zea mays*. *Planta*, 81, 28.

Rebhun, L. I. (1972). Freeze-substitution and freeze-drying. *In:* Principles and Techniques of Electron Microscopy: Biological Applications, Vol. 2 (Hayat, M. A., Ed.), pp. 3–52. Van Nostrand Reinhold Company, New York.

Remagen, W., Höhling, H. J., Hall, T. A., and Caeser, R. (1969). Electron microscopical and microprobe observations on the cell sheath of stimulated osteocytes. *Calc. Tissue Res.*, 4, 60.

Robison, W. L., and Davis, D. (1969). Determination of iodine concentration and distribution in rat thyroid follicles by electron probe microanalysis. *J. Cell Biol.*, 43, 115.

von Rosenstiel, A. P., and Zeedijk, H. B. (1968). An electron probe and electron microscope investigation of asbestos bodies in lung sputum. *Proc. III Natl. Conf. Electron Probe Analysis*, Chicago, Ill., p. 43A.

Roth, L. J., and Stumpf, W. E. (1969). Autoradiography of diffusible substances. Academic Press, New York, 371 pp.

de Rouffignac, C., Lechène, C., Guinnebault, M., and Morel, F. (1969). Etude par microponction de l'élaboration de l'urine. III. Chez le Mérion non diurétique et en diurèse par le mannitol. *Nephron*, 6, 643.

Russ, J. C. (1971). Energy dispersion X-ray analysis on the scanning electron microscope. *In:* Energy Dispersion X-ray Analysis: X-ray and Electron Probe Analysis (Russ, J. C., Ed.), pp. 154–179. Amer. Soc. Testing and Materials, Philadelphia, Penna.

Sakai, M., Austin, J., Witner, F., and Trueb, L. (1969). Studies of corpora amylacea. I. Isolation and preliminary characterization by chemical and histochemical techniques. *Arch. Neurol.*, 21, 526.

Schmitt, W. W., Zingsheim, P., and Bachmann, L. (1970). Investigation of molecular and micellar solutions by freeze-etching. *VII. Intl. Conf. Electron Microsc., Grenoble,* Vol. I, p. 455.

Sims, R. T., and Hall, T. A. (1968). X-ray emission microanalysis of proteins and sulfur in rat plantar epidermis. *J. Cell Sci.,* **3,** 563.

Spurr, A .R. (1969). A low-viscosity epoxy resin embedding medium for electron microscopy. *J. Ultrastruct. Res.,* **26,** 31.

Staehelin, L. A., and Bertaud, W. S. (1971). Temperature and contamination dependent freeze-etch images of frozen water and glycerol solutions. *J. Ultrastruct. Res.,* **37,** 146.

Stenn, K., and Bahr, G. F. (1970a). Specimen damage caused by the beam of the transmission electron microscope. *J. Ultrastruct. Res.,* **31,** 526.

———— (1970b). A study of mass loss and product formation after irradiation of some dry amino acids, peptides, polypetides, and proteins with an electron beam of low current density. *J. Histochem. Cytochem.,* **18,** 574.

Stirling, C. E., and Kinter, W. B. (1967). High resolution radioautography of galactose-^{3}H accumulation in rings of hamster intestine. *J. Cell Biol.,* **35,** 585.

Stuve, J. and Galle, P. (1970). Role of mitochondria in the handling of gold by the kidney. A study by electron microscopy and electron probe analysis. *J. Cell Biol.,* **44,** 667.

Sutfin, L. V., Holtrop, M. E., and Ogilvie, R. E. (1971). Microanalysis of individual mitochondrial granules with diameters less than 1000 Å. *Science,* **174,** 947.

Tandler, C. J., Libanati, C. M., and Sanches, C. A. (1970). The intracellular localization of inorganic cations with potassium pyroantimonate. Electron microscope and microprobe analysis. *J. Cell Biol.,* **45,** 355.

Thomas, R. S. (1969). Microincineration techniques for electron microscopic localization of biological materials. *In:* Advances in Optical and Electron Microscopy, Vol. III (Barer, R., and Cosslett, V. E., Eds.), pp. 99–150. Academic Press, London.

Wachtel, A. W., Gettner, M. E., and Ornstein, L. (1966). Microtomy. *In:* Physical Techniques in Biological Research, Vol. 3 (Pollester, A. W., Ed.), pp. 173–250. Academic Press, New York.

Waisel, Y., Hoffen, A., and Eshel, E. (1970). The localization of aluminum in the cortex cells of bean and barley roots by X-ray microanalysis. *Physiol. Plant.,* **23,** 75.

Warner, R. R. (1972). Application of the electron probe microanalyzer to the quantitative analysis of biological material and the analysis of intestinal calcium transport. Dissertation, Univ. of Rochester, 338 pp.

————, and Coleman, J. R. (1972a). A computer program for quantitative microanalysis of thin biological material. *Proc. VII Natl. Conf. Electron Probe Analysis, San Francisco, Calif.,* 41A.

————, and Coleman, J. R. (1972b). Calcium transport in the small intestine. *Proc. VII Natl. Conf. Electron Probe Analysis, San Francisco, California,* p. 44A.

————, and Coleman, J. R. (1973). A quantitative procedure for electron probe analysis of biological material. *Micron* **4**, 61.

Watson, M. L., and Aldridge, W. G. (1961). Methods for the use of indium as an electron stain for nucleic acids. *J. Biophys. Biochem. Cytol.*, **10**, 257.

Wolstenholme, G. E. W., and O'Connor, M. (Eds.) (1969). The Frozen Cell, CIBA Foundation Symp. Little, Brown and Company, London and New York.

Zingsheim, H. P. (1972). Membrane structure and electron microscopy. The significance of physical problems and techniques. (Freeze-etching.) *Biochem. Biophys. Acta*, **265**, 339.

Note: Proceedings of the national conferences on electron probe analysis may be obtained from: Dr. Joseph I. Goldstein. Metallurgy and Materials Science Department, Lehigh University, Bethlehem, Penna.

AUTHOR INDEX

Warner, R. R., 163, 164, 167, 176, 178, 190, 193, 196, 201, 206
Watson, D. H., 92, 96, 97, 121, 122, 124, 128
Watson, M. L., 177, 207
Wensink, P. C., 65, 70, 74, 80
Westmoreland, B. C., 68, 73, 80
Wildy, P., 97, 128
Williams, R. C., 90–93, 124, 126, 128, 132
Williams, W. C., 111, 126
Wilson, C., 186, 204
Wilson, H. B., 179, 204
Witner, F., 187, 205
Wittkopp, R. W., 198, 204
Wittry, D. B., 171, 204

Wolfson, 65, 70, 72, 79
Wolstenholme, G. E. W., 172, 207
Woolgar, A. E., 174, 186, 202
Wrigley, R. C., 143, 156
Wyckoff, R. W. G., 131, 155
Wynne-Jones, W. F. K., 143, 157

Yen, S.-S., 188, 203
Yoshiike, K., 70, 80
Yuile, C. L., 179, 204

Zahn, R. K., 66, 67, 80
Zeedijk, H. B., 180, 205
Zeitler, E., 16, 36–39, 44, 177, 200
Zingsheim, P., 174, 205, 207
Zirkin, B. R., 170, 203

SUBJECT INDEX

Acrolein, 168, 190
Albumin, 95
Alcian blue, 55, 60
Ammonium bicarbonate, 136
Amyl acetate, 57, 95, 101
Artifact, 67, 141, 167, 174, 175, 179, 187, 190
Asbestos, 179, 180, 182, 183
Astigmatism, 5, 40
Atomizer, 93, 94
Azure II, 55, 60

Back focal plane, 2, 3
Bacteria, 45, 131
Bacterial spores, 130
Barium, 188
Blood, 144, 192, 193
Bone, 131, 165, 170, 183, 184
Brain, 186, 187, 197, 198
Buffer, 53, 54

^{14}C-acetate, 175
Calcium, 150, 153, 160, 161, 166–169, 171, 178, 179, 184, 185, 187–191, 193, 196
^{40}Ca, 166
^{45}Ca, 166, 167
Calcium oxalate, 172
Calcium phosphate, 136, 145, 167, 185
Carbon layer, 48, 49, 53–56
Carbon support film, 7, 22, 24, 33, 39, 67–69, 83, 87, 106, 132, 133, 182
Cardiac muscle, 81, 82
Cartilage, 184
Centrifuge cells, 104–106, 108, 109, 111
Charging, 179
Chlorine, 161, 175, 186, 192
Collagen, 9, 11, 47, 138, 139, 150, 152
Collodion, 98, 103, 133
Computer, 17, 21, 23, 42, 43, 69, 70
Condenser lens, 3, 4
Conical illumination, 18, 19, 27, 32
Contamination, 33–36, 38, 39, 43, 141, 145, 161, 182, 193
Copper, 161, 192
Critical-point drying, 55, 57, 58, 61, 170
Crossover, 3, 5, 11, 31
Crystal lattices, 2

Damage, 178, 179
Darkfield, 5, 6, 9, 13, 16
Densitometer, 17
Diffraction grating replica, 92
Diffraction mode, 5
Diffraction pattern, 2, 13
Diffusion, 176, 177, 189, 193
Distortion, 144, 145, 150, 171, 176, 184

E. coli, 73, 76
Electron paramagnetic resonance, 194
Epithelial cells, 149, 150
Epoxy resins, 47, 48
Erythrocytes, 114, 115, 166, 174, 191
Europium, 179, 181

Faraday cage, 17
Ferritin, 174
Flame photometry, 186
Fogging, 20, 38
Formaldehyde, 66, 67, 95, 179, 180, 184, 186–188, 190, 197
Formamide, 66, 68
Formvar, 24, 56, 83, 96, 97, 102, 133, 134
Foucault contrast, 1
Fourier theory, 1
Freon, 57, 186
Fresnel fringes, 1

Gelatin, 167, 174, 187
Glutaraldehyde, 53, 115, 135, 136, 138, 139, 141, 145, 148–150, 152–154, 161, 172, 186, 189, 192
Glycerol, 174
Glycogen, 175
Gold, 179, 180

Halogen, 178
Heme, 165
High resolution, 16, 27, 39, 46, 59, 135
Histogram, 70, 72
HPMA, 47

Indium, 177
Interference colors, 134
Intestinal epithelium, 194
Intestine, 131, 190
Iodine, 144, 186, 187
Iron, 161, 165